THE McGRAW-HILL CIVIL PE EXAM DEPTH GUIDE
Transportation Engineering

THE McGRAW-HILL CIVIL ENGINEERING PE EXAM DEPTH GUIDE
Transportation Engineering

James T. Ball, PE

McGRAW-HILL
New York Chicago San Francisco Lisbon London Madrid
Mexico City Milan New Delhi San Juan Seoul
Singapore Sydney Toronto

Cataloging-in-Publication Data is on file with the Library of Congress

McGraw-Hill
A Division of The McGraw·Hill Companies

Copyright © 2001 by The McGraw-Hill Companies, Inc. All rights reserved. Printed in the United States of America. Except as permitted under the United States Copyright Act of 1976, no part of this publication may be reproduced or distributed in any form or by any means, or stored in a data base or retrieval system, without the prior written permission of the publisher.

1 2 3 4 5 6 7 8 9 0   AGM/AGM   0 7 6 5 4 3 2 1

ISBN 0-07-136180-4

The sponsoring editor for this book was Larry S. Hager and the production supervisor was Sherri Souffrance. It was set in Times Roman by Lone Wolf Enterprises, Ltd.

Printed and bound by Quebecor/Martinsburg.

 This book is printed on recycled, acid-free paper containing a minimum of 50% recycled, de-inked fiber.

McGraw-Hill books are available at special quantity discounts to use as premiums and sales promotions, or for use in corporate training programs. For more information, please write to the Director of Special Sales, McGraw-Hill, Professional Publishing, Two Penn Plaza, New York, NY 10121-2298. Or contact your local bookstore.

Information contained in this work has been obtained by The McGraw-Hill Companies, Inc. ("McGraw-Hill") from sources believed to be reliable. However, neither McGraw-Hill nor its authors guarantee the accuracy or completeness of any information published herein, and neither McGraw-Hill nor its authors shall be responsible for any errors, omissions, or damages arising out of use of this information. This work is published with the understanding that McGraw-Hill and its authors are supplying information but are not attempting to render engineering or other professional services. If such services are required, the assistance of an appropriate professional should be sought.

CONTENTS

Introduction   x
About the Author   xiii

CHAPTER 1: HORIZONTAL AND VERTICAL ALIGNMENT   1.1

Tangents and Horizontal Curves   1.1
Spiraled Railroad Curves   1.4
Superelevation   1.4
Vertical Alignment and Design Speeds   1.6
Vertical Curves and Sight Distance   1.7
Example Problem 1.1   1.10

CHAPTER 2: EARTHWORK   2.1

The Earthwork Profile and Mass Diagram   2.1
Cross-Sections: Cuts and Fills   2.2
Waste and Borrow   2.2
Haul   2.3
Example Problem 2.1   2.3

CHAPTER 3: DRAINAGE   3.1

Curbs and Gutters   3.1
Ditches and Edge Drains   3.2
Culverts   3.3
Precast Culverts and Small Bridges   3.4

CHAPTER 4: PROTECTION OF THE ENVIRONMENT   4.1

Avoidance of Environmental Damage   4.1
Noise Pollution   4.1

Historical Sites, Public Buildings, and Parks 4.2
Wetlands Filling 4.2
Erosion and Siltation 4.2
Air Pollution 4.3
Wildlife Habitat 4.4
Example Problems 4.4

CHAPTER 5: ROADWAY DESIGN 5.1

Design Vehicles and Turning Widths 5.1
Lanes 5.3
Shoulders and Ditches 5.4
Medians 5.4

CHAPTER 6: TRAFFIC ESTIMATION AND DESIGN VOLUMES 6.1

Traffic Counting and Annual Increase 6.1
Traffic Forecasts 6.1
Lane Capacity 6.2
Lanes Needed at Design Year 6.4

CHAPTER 7: INTERSECTION DESIGN 7.1

Low-Volume Intersections 7.1
Volume-Modified Intersections 7.3
Indirect Left Turns and U-Turns 7.5
Roundabouts and Center Turning Overpass 7.5

CHAPTER 8: FREEWAY AND INTERCHANGE DESIGN 8.1

Freeway and Interchange Warrants 8.1
Freeway Design 8.2
Grade Separations 8.3
Diamond Interchanges 8.4
Modified Diamonds 8.6
Cloverleaf Interchanges 8.8
Directional Interchanges 8.12

CHAPTER 9: PAVEMENT BASE   9.1

    Suitable Subbase Material 9.1
    Pozzolan and Asphalt Amendments 9.5
    Portland Cement Concrete Base 9.5

CHAPTER 10: RIGID PAVEMENT DESIGNS   10.1

    Pavement Design and Construction 10.1
    Mix Designs 10.2
    Continuous Slab 10.4
    Prestressed Concrete Pavement 10.4

CHAPTER 11: FLEXIBLE PAVEMENT DESIGN   11.1

    Subgrade and Subbase Construction 11.1
    Base and Pavement Constructon 11.2
    Pavement Reconstruction 11.3
    Recycled Asphalt 11.4
    Asphalt Cements and Superpave Mixes 11.5
    Additives and Rubber Crumbs 11.6
    Stone Matrix Asphalt 11.7

CHAPTER 12: AIRPORT ENGINEERING   12.1

    Airport Access 12.1
    Runway Configurations 12.2
    Runway Grades and Slopes 12.2
    Flexible Pavements 12.4
    Rigid Pavements 12.7

CHAPTER 13: RETAINING WALLS   13.1

    Choosing Retaining Walls 13.1
    Mechanically Stabilized Backfill and Reinforced Earth 13.2
    Gravity Walls 13.2
    Cantilevered and Counterfort Walls 13.4

CHAPTER 14: HIGHWAY-RAILROAD CROSSINGS 14.1

Rural Crossings 14.1
Urban Crossings 14.3
Separations 14.4

CHAPTER 15: BICYCLE FACILITIES 15.1

Bikeways 15.1
Bike Lanes 15.2
Bicycle Streets 15.2
Urban Congestion Solutions 15.3

CHAPTER 16: TRANSIT 16.1

HOV Lanes 16.2
Bus Transit 16.2
Light Rail 16.4
Heavy Rail 16.4
Commuter Trains 16.6

CHAPTER 17: RAILWAY ENGINEERING 17.1

Construction 17.1
Efficiency of Operation 17.3
Passenger Capacity 17.3
High-Speed Rail 17.4
Maglev 17.7

CHAPTER 18: PEDESTRIAN FACILITIES 18.1

Sidewalks 18.1
Street Crossings 18.3
Pedestrian Bridges 18.4

CHAPTER 19: RIGHT-OF-WAY 19.1

Right-Of-Way Planning 19.1
Scheduling 19.2

Temporary Construction Easements 19.2
Permanent Road Easements 19.3
Joint Use 19.3

CHAPTER 20: TRAFFIC MANAGEMENT 20.1

Auxiliary Lanes 20.1
Signing and Marking 20.4
Signal Lights 20.5
Light Coordination 20.5
Traffic Calming 20.6
Parking Control 20.6
Maximizing Arterial Capacity 20.7

APPENDIX A: BRIDGE ENGINEERING A.1

APPENDIX B: AIRPORT ENGINEERING B.1

APPENDIX C: HIGHWAY ENGINEERING C.1

INDEX I.1

INTRODUCTION

Transportation Engineering includes planning, design, and supervision of the construction, maintenance, and operational phases of the nation's infrastructure which is related to the movement of people and goods. The primary industries involved are roads, railroads, air transport, and city transit. Although transportation engineering training and experience are applicable to more than one industry, few engineers choose to work in more than one.

The trend for decades, if not centuries, has been toward specialization, under the impetus of increasing knowledge, new materials, and technologies. Within each industry, most engineers are now specialists rather than generalists, although many change specialties during their typically long transportation engineering careers. Among the hundreds of specialties in the field are highway design, highway or railroad location, bridge design, construction inspection, airport runway pavement design, county or low-volume road design, drainage analysis and design, urban traffic management, highway maintenance, and transit planning.

Transportation is a fluid field. Many specialties are evolving into subspecialties. Specialists must become generalists, however, as they enter engineering management and assume responsibility for many specialties. Much of the work is done by technicians and other subprofessional personnel, and by young engineers in training, who should be under the supervision of professional engineers. Professional licensing, in the form of the Professional Engineering (PE) certification, recognizes the technical qualifications of engineers by their peers, other professionals, and the government.

Problems, such as early material failure and high accident experience, arise in the field and are solved in the field; sometimes in innovative ways. These field innovations, as well as the results of continual research and experimentation, are reported in technical journals. Scientists develop new materials and processes. Engineers make incremental improvements in the operation and economics of vehicles, pavements, bridges, rails, and fuels. This gives us continual improvement in technology, such as stronger and cheaper pavements, longer bridge spans, containerization of freight and double-stacking of containers, and advances in diesel and electric propulsion.

The American transportation system is admired and emulated throughout the world. Inexpensive travel and freight contribute to a historically high standard of living in North America, Europe, and Japan—and other areas of the globe are racing to catch up. Many problems remain to be solved, some of them increasing in intensity while awaiting solutions. Transportation engineers are in the forefront of this work, acting directly or recommending political action.

Heavy trucks require much greater stopping distance than automobiles. Auto drivers, closer to the ground, require more control of grade differences on summit curves. Trucks require more downgrade control, heavier pavements and bridges, and careful attention to horizontal curve superelevations on upgrades. As trucks become longer, wider, heavier, and more numerous in the traffic stream, the difference between roadway requirements for trucks and cars is more critical. Large trucks are essential to freight economics and the increase in truckload limits might be important to the national economy, but those trucks also can endanger the lives and property of those in automobiles.

All highway design involves trade-offs. How much safety can be provided on a specific highway? Will costs unduly delay other urgently needed projects? In some cases, we might reach a point where providing a separate highway for trucks is preferable to the maintenance and reconstruction costs of repeated heavy axle loadings on multi-lane highways. Truck weight and size limits and restrictions to certain hours, and numerous other alternatives are in place or are under consideration because of the mixed traffic. Another concern is how to reduce the 5,000 to 6,000 fatalities that occur annually in auto/truck accidents, of which 97 percent of the victims are drivers and passengers of autos.

Few of us still believe that, within current technology constraints, we can build sufficient urban lanes for commuters. Improved transit and traffic management within the cities can help. Perhaps smaller urban vehicles, allowing narrower lanes, and intelligently designed highways that safely allow lesser headways, can help us keep up; but we now are falling behind rapidly in critical cities such as Houston, Los Angeles, Washington, New York, and Seattle. The situation seems ripe for changes that periodically revolutionize our profession.

Within our nation's history, we have experienced successive worldwide transportation revolutions: Steamships, railroads, city transit, automobiles, long-span bridges, airplanes, jet engines, interstate highways, and tunnel boring. Europe and Japan recently have led in high-speed train and magnetic levitation technology that might be important to us and to other nations with long distances between cities.

Other revolutions in transportation are certain to occur but are not yet fully envisioned. A possible scenario is the development of magnetic levitation. After the initial testing period, engineers will work to solve speed restrictions caused by air resistance. This could lead to encasement within an evacuated tube, with minimal and gradual alignment or grade changes. Such design will require higher bridges and tunneling under mountains—a stretch of current technology. With no friction and greatly reduced air resistance, speeds of Mach 2, 3, or more might be attainable, depending on the degree of vacuum in the tube and the degree of control of the gradient. A transcontinental surface vehicle might take less time than a supersonic aircraft and not cause stratospheric pollution. In fact, our air pollution problems could be reduced with the solution of the clean production of electric power.

Could an evacuated tube be anchored below wave action and shipping depth at Gibraltar? The Dardanelles? Bering Strait? The English Channel? The Kattegat? Puget Sound? San Francisco and Chesapeake bays? Successful construction of such tubes might lead to the elimination of many ferries and many transoceanic aircraft flights.

Can the technology and economics be attained (on land if not across the waters), with acceptable compromises on design speed? If so, these facilities will be the solution to our continuing escalation of airport and airway congestion. Much of our interurban highway congestion, transportation accident carnage, and air pollution could be reduced, while increasing travel comfort and fractionalizing travel time. If magnetic levitation technology can be made viable for freight, many issues could be addressed, enhancing express freight and just-in-time deliveries, reducing freight costs, air pollution, and auto/truck highway accidents.

Solutions to urban traffic congestion also will require innovation. We have found that our city centers are too valuable and convenient to abandon. We might implement "back-to-the-future" actions to make the inner cities as livable as they once were, or more so, despite increased population density. Transportation engineers have a duty to offer advice and guidance to political leaders who must make tough transportation decisions.

Viable long-range urban solutions currently involve the redirecting and calming of auto traffic. Redirection would involve improved arterials leading directly to parking facilities with a transit link at the city core boundary. Within that core—essentially reserved for pedestrians—the only large vehicles permitted would be emergency vehicles and late-night supply trucks. Substitution for the restricted traffic in the city core would involve 24-hour transit service, with streets reserved for bicycles, skates, walking, and small vehicles powered by electricity. Bus service would be provided in a loop around the city core.

Traffic redirection encourages using the total width of arterials for through traffic, prohibiting parking and left turning, timing traffic lights to facilitate the main traffic direction with little stopping, and providing separations for the major cross traffic. Traffic calming methods, developed to discourage use of residential and other streets by through traffic, are traffic circles, speed humps, serpentines, jogs, and narrowing.

The future in transportation is left to the vision and talents of the civil engineers now taking the PE examination. Welcome to our challenging profession!

James T. Ball, PE

ABOUT THE AUTHOR

JAMES T. BALL, PE

James T. Ball, PE, is former director of the Transportation Division of the Bureau of Indian Affairs, Department of the Interior. A practicing civil engineer for 36 years, during a career of road and highway planning, location, and construction, he has overseen the transport of numerous structures out of highway right of ways, and the rehabilitation of many structurally threatened historic buildings.

CHAPTER 1
HORIZONTAL AND VERTICAL ALIGNMENT

TANGENTS AND HORIZONTAL CURVES

Designers try to make tangents and the curves between them as long as it is economically viable to make the roadway safer and more comfortable. In rolling or difficult terrain, vertical tangents and curves are usually much shorter. Small changes of horizontal alignment can be made at angle points without curvature at the intersection when a curve can be accomplished within the driver's lane without difficulty.

The mathematics of tangents and curves are relatively simple trigonometry, geometry and algebra, and the combined use of them. On the right triangle shown in Figure 1.1, sides a, b, and c respectively are opposite angles A, B, and C.

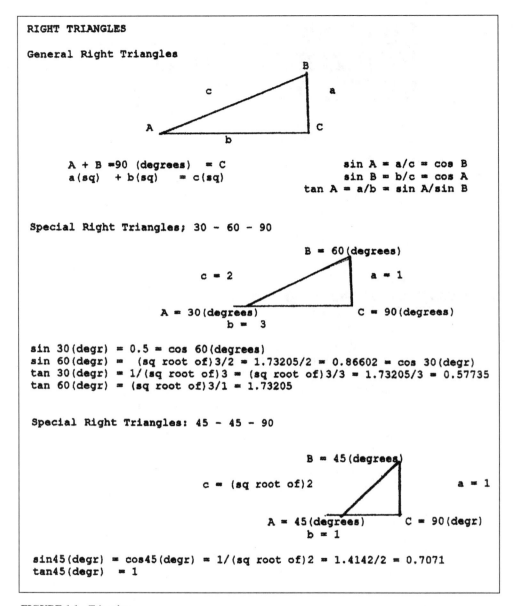

FIGURE 1.1 Triangles.

Angle C is the right angle, and c is the hypotenuse of the triangle.
Angle A + Angle B = Angle C = 90°.
Sin A = a/c   sin B = b/c = sin (90° − A)   cos A = b/c
cos B = a/c = sin A   tan = a/b = sin A/sin B   tan B = b/a = sin B/sin A
In a 30-60-90 triangle, sin 30° = 0.5
In an isosceles 45-45-90 triangle, tan 45° = 1

The special case triangles—30-60-90 and 45-45-90—are convenient to simplify work, especially if the survey instrument cannot occupy the point of intersection (PI) of the tangents of a curve.

The square of the hypotenuse of a right triangle is equal to the sum of the squares of the other two sides. $c^2 = a^2 + b^2$.

On a circle, the circumference = 2 × radius × π, with π approximately 3.1416.

On a circular curve, the curve distance is 2 × radius × π × central angle × $\frac{1}{360}$.

The central angle of the curve is equal to the change of angle at the PI between the two tangents, as shown in Figure 1.2.

With these simple mathematical facts and their derivations, and trigonometrical tables, a calculator, or a slide rule (remember those?) almost any tangential and horizontal curve problem can be solved.

```
Angle BAC = Angle CAD = Angle EBC = Angle EDC = 90(degr) - AngleBEC

Angle BEC = Angle ABC = Angle ADC = Angle CED = 90(degr) - Angle BAC

Angle FED = 180(degr) - 2(Angle BEC) = 2(Angle BAC)= Central angle
(delta)

Long Chord BD = 2r sin (delta)/2.  Arc Distance BD =(delta)/360 x (pi)d

Tangent Distance BE = r tan (delta)/2.

The angle between the tangent BE and the chord BD; that is, between the
PI and PT as measured at the PC; is one-half of the arc BD, or one half
of the central angle (delta).  Similarly, a station 50m from the PC
would be at an angle of one half of 360 x 50/2(pi)r measured at the PC
from the tangent.  The angle at the next 50 meter station would be 1/2
x 360 x 100/2(pi)r, and so on until the stationing total equals the arc
distance, at the PT.  All stations can thus be run in from the PC,
measuring along the curve.  The difference between arc and chord
distances between stations along the curve may be disregarded when
deflection angles are less than 5(degrees).  For larger deflection
angles, measure each chord from the PC, using the formula:  chord
distance = 2 r sin 1/2 (delta).
```

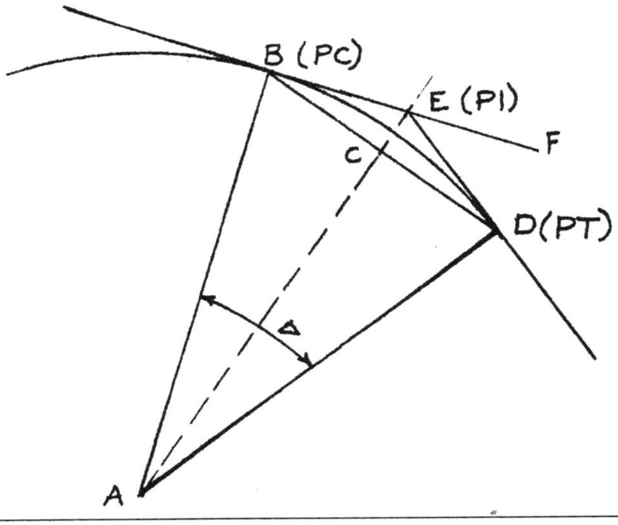

FIGURE 1.2 Arcs, tangents, and chords.

SPIRALED RAILROAD CURVES

Horizontal curves are circular, except for railroad curves that usually are spiraled at the tangents. A highway tangent to curve would be spiraled only with a combination of high speed and sharp curvature, and the usual practice is to flatten the curve, permitting the vehicle to follow a spiral within the lane width when the driver turns the steering wheel gradually to move the car from the tangent to the curve.

In 1909, W. H. Short of the London Institution of Civil Engineering worked out the formula still used for the length of railroad curves by gradual gain of centripetal acceleration:

$$L = 0.0702 V^3/RC$$

where L = Minimum length for the spiral in meters
  V = Velocity in km/h
  R = Radius of the curve in meters
  C = Factor based upon the desired safety and comfort. $C = 1$ is usually used in railroad design, and $C = 1, 2,$ or 3 in highway design.

Since highways require less precise spirals, instead of the above formula, the spiral length usually is made to match that of the superelevation runoff, with the exception of short radius curves such as those used in interchange design.

A spiral's radius varies inversely with the distance along the spiral. In the case of a spiral from a tangent to a curve, the spiral radius varies from the infinite radius of the tangent to the radius of the curve. In the case of a spiral used to ease the transition between two curves, the spiral radius varies from the radius of one curve to the radius of the other. Spirals are the best way to make the transition between reverse curves or between compound curves (curves of considerably different radii in the same direction), or to replace the tangent on broken-back curves (curves in the same direction with a short tangent between them).

SUPERELEVATION

Superelevation is the elevation on curves of the outside edge of the pavement above the centerline and of the inside edge of the pavement below the centerline. Superelevation runoff length usually is set to be aesthetically pleasing to the driver, or somewhat longer than would be required operationally. For the highest speeds, above 80 km/h, the greatest grade differential is 1:200; that is, the difference between the longitudinal grade at the centerline and the longitudinal grade at the pavement edge should not exceed 0.5 percent. For lower speeds, a greater grade difference with a shorter runoff can be used. Using these criteria, runoff distance in meters for two lanes has been tabulated as shown in Table 1.1.

TABLE 1.1 Superelevation Runoff (two lanes)

Superelev %	Design Speed km/hr									
	30	40	50	60	70	80	90	100	110	120
2	20	25	30	35	40	50	55	60	65	70
4	20	25	30	35	40	50	55	60	65	70
6	30	35	35	40	40	50	55	60	65	70
8	40	45	45	50	55	60	60	65	70	75
10	50	55	55	60	65	75	75	80	85	90
12	60	65	65	75	80	90	90	95	105	110

For wider undivided highways, use:

Three-lane highways	Tabular value × 1.2
Four-lane highways	Tabular value × 1.5
Five-lane highways	Tabular value × 1.8
Six-lane highways	Tabular value × 2.0.

For the superelevation itself, formulae have been developed involving the side friction of tires and the speed of the vehicle, but so many other factors are involved, including inclement weather, tire condition, road surface texture, and vehicular differences, that the maximum gradients commonly used are those shown in Table 1.2.

TABLE 1.2 Superelevations

Location	Recommended Maximum Superelevation
Open rural highways without snow and ice	0.12
Urban highways without snow and ice	0.08
Low-volume gravel highways	0.08
Open rural highways subject to snow and ice	0.08
Urban highways subject to snow and ice	0.06
Where traffic congestion or other conditions slow traffic	0.04

Remember, these are maximum gradients. Wherever weather or traffic conditions are likely to be extreme, superelevation should be lower. Special circumstances also must be considered, such as in the case of a long, sustained upgrade followed by a superelevated curve that is not an obstacle to automobiles, but might cause trucks with a high center of gravity to enter the curve slowly enough to lose part of a load, or even to overturn.

VERTICAL ALIGNMENT AND DESIGN SPEEDS

Long tangents, especially tangents connected by long curves, are desirable in vertical as well as in horizontal alignment. They are, however, much more difficult to achieve, especially in rolling or mountainous terrain. Automobiles exhibit little loss of speed up to 3 percent sustained grades, and all but a few maintain their speed up to 5 percent. Speed differences between autos on upgrades and downgrades below 5 percent are moderate. Trucks and automobiles tend to operate near the same range on level and near-level grades.

Trucks, however, are much more affected by sustained upgrades above 3.5 percent, and downhill and uphill speeds are about 6 percent higher and lower respectively. The characteristics of recreational vehicles vary widely, but the downhill and uphill speeds are generally between those of automobiles and trucks.

From these criteria, a maximum upgrade desirable for least interference with mixed vehicular operation would be 3.5 percent, or 3 percent for very long sustained upgrades. If these grades were exceeded, lower superelevation rates would make the curves safer for trucks, but cars then would be less safe if the speed limit were not reduced.

Speeds on urban streets are controlled by speed limits and traffic conditions, making capacity more important than high-speed design. Design speed usually is meant for rural highways and urban freeways, and it is chosen as the highest reasonable speed attainable under prevailing conditions. Highways with local conditions that make design speed unattainable should be engineered for maximum driver safety and driver awareness of necessary speed reductions. The selected design speed establishes the minimum curve radius, minimum sight distance, and maximum grade, as shown in Table 1.3.

TABLE 1.3 Recommended Maximum Grades

Design Speed, km/hr	Maximum Grade, %
110	5
100	5.5
90	6
80	6.5
70	7
60	7.5
50 (Important highways)	8
50 (Low-volume highways)	12

Exceptions:
Short grades less than 150 meters, and one-way downgrades, may be 1% steeper.
Low-volume roads may be 2% steeper.

The design grade chosen should be the least one that is economically feasible. The minimum recommended grade is 0.5 percent, which facilitates drainage of curbed highways and streets, and applies to the gradient of rural drainage ditches and urban gutters.

When the maximum grade shown by Table 1.5 must be exceeded, other design options should be investigated, including lower design speed and a lower posted speed limit.

VERTICAL CURVES AND SIGHT DISTANCE

For stopping sight distance on summit vertical curves, the standard height of the eye of the driver of an automobile is 1070 mm above the roadway, and the top of the object observed is 150 mm above the roadway. The basic formulae for the length of a parabolic crest vertical curve for that stopping sight distance were based upon the algebraic difference of grades, sight distance, and an empirical factor.

Although they are basic for design, transportation engineers now use these formulae as modified by other controls. With sight distance S less than length of curve L, and A the difference of grades (% grade change):

$$L = S^2A/404; \text{ and with } S > L, L = 2S - 404/A. \text{ If } L = S, L = 404/A$$

The values given by these formulae do not always give reasonable solutions for short curves and do not represent current practice. Most states have established a minimum crest vertical curve length based on a selected value, or several lengths based upon a range of speeds, or a length based upon the change of grade. Recognizing that a vertical curve length well above minimum is desirable, the AASHTO approximation of the median of these current state practices for determining the minimum length of a crest vertical curve with length less than 100 m is 0.6 times the design speed, or

Recommended Minimum L = 0.6V, with L in meters and V in km/h.

Another design control, valid over the range of crest vertical curves greater than 100m, is minimum L = KA (Table 1.4). By this control, a vertical curve with a 3 percent change of grade and an 80 km/h design speed would have a minimum length of 49 × 3 m = 147m. Using the L = 0.6D control, L = 0.6 × 80 m = 48 m minimum.

TABLE 1.4 Design Controls for Crest Vertical Curves

Design Speed Km/h	Coefficient of Friction (f)	Rate of Vertical Curvature K, m/% of A
30	0.40	3
40	0.38	5
50	0.35	10
60	0.33	18
70	0.31	31
80	0.30	49
90	0.30	71
100	0.29	105
110	0.28	151
120	0.28	202

These k values may be used without appreciable error, with S greater than or less than L.

$k = 51$ is the recommended maximum for drainage. It can be exceeded, but with special attention to drainage near the apex of the crest vertical curve.

Stopping sight distance studies for sag vertical curves assume a headlight beam angled upward one degree and at 600 mm above the pavement, lighting the surface of the roadway. Minimum length of curve L is in m, stopping sight distance S is in m, and difference of grade A is described in percentages.

In sag vertical curves, it is recommended that for road user comfort a much longer curve than the minimum for stopping sight distance be designed for the design speeds up to 80 km/h. A moderately longer curve than the minimum can be planned at design speeds from 80 to 100 km/h. At a design speed greater than 100 km/h, a k value of 51 will be reached, which is the maximum recommended for drainage. When k exceeds 51, the minimum curve length for stopping sight distance can be used, but special attention should be given to drainage near the low point of the curve.

For sag vertical curves:
When S < L, Recommended L = $S^2A/(120 + 3.5S)$
When S > L, Recommended L = $2S - (120 + 3.5S)/A$
When S = L, L = $120/(A - 3.5)$.

where L = Curve length in meters
       S = Light beam distance in meters
       A = Change of grade in percent (Table 1.5).

TABLE 1.5 Design Controls for Sag Vertical Curves

Design Speed Km/h	Coefficient of Friction (f)	Rate of Vertical Curvature K m per % of A
30	0.40	4
40	0.38	8
50	0.35	12
60	0.33	18
70	0.31	25
80	0.30	32
90	0.30	40
100	0.29	51
110	0.28	62
120	0.28	73

For a 6 percent change of grade, when design speed is 100 km/h

L = (2) (205) − 1\6 [120 + (3.5) (205)] = 270m minimum, S < L

Horizontal curve stopping sight distance, as compared to vertical curve stopping sight distance, is within the control of the designer and the maintenance engineer. Improvement of horizontal stopping sight distance usually means removing visual obstructions on the inside of curves; cutting vegetation or cutting back a cut slope and widening the shoulder and ditch. The designer can move the road away from a vision-obstructing cut slope by lengthening a horizontal curve. Where this is impractical, a lower speed limit with warning signs is necessary if the design speed exceeds the safe stopping speed.

Stopping sight distance is composed of the distance traveled during reaction time, which varies widely among individuals and for design purposes is usually set at 2.5 seconds, and at the braking distance. Safe stopping distance is usually calculated for a wet pavement. It is not practical to design highways for high speed on wet pavement, and drivers must be taught to slow for wet pavement and any other condition affecting the friction between the tires and the pavement. Stopping sight distance is approximated by the standard formulas:

$$d = V^2/254f \text{ and } d' = 0.694V$$

where d = Braking distance in meters
V = Initial speed in km/h
f = Coefficient of friction between the tires and roadway (Table 1.6)
d' = Reaction distance in meters

TABLE 1.6 Stopping Sight Distance, Wet Pavements

Design Speed Km/h	Reaction at 2.5 s	f	Braking Distance m	Stopping Sight Distance, m
30	20.8	0.40	8.8	29.6
40	27.8	0.38	16.6	44.4
50	34.7	0.35	28.1	62.8
60	41.7	0.33	42.9	84.6
70	48.6	0.31	62.2	110.8
80	55.5	0.30	84.0	139.5
90	62.5	0.30	106.3	168.8
100	69.4	0.29	135.8	205.2
110	76.4	0.28	170.1	246.5
120	83.3	0.28	202.5	285.8

These stopping sight distances are for a zero gradient; downgrades require more distance and upgrades less, approximately 1% distance adjustment per percent of gradient at 45 km/h, 2% at 100 km/h, and 3% at 130 km/h.

Passing sight distance usually is not provided on crest vertical curves or on city streets, and it is seldom a problem on multi-lane facilities. Two-lane rural highways and some suburban highways should provide passing opportunities at frequent intervals and for substantial portions of their length. Where traffic warrants it, and where few such passing opportunities exist, passing lanes should be provided.

In 1941, C.W. Prisk investigated passing practices on rural two-lane highways. Later studies have confirmed that his data are accurate and conservative for modern vehicles. He found the passing vehicle averaged 15 km/h more than the overtaken vehicle, and he measured for each of several representative passing speeds these four values:

1. The time and distance of the initial maneuver to the lane change
2. The time and distance of the period occupying the opposing lane
3. The distance of clearance of opposing traffic after the passing
4. The distance covered by an opposing vehicle during two-thirds of the time the opposing lane was occupied during the passing. (Table 1.7)

TABLE 1.7 Safe Passing Distance

Average Passing Speed, km/hr	56.2	70.0	84.5	99.8
Initial maneuver:				
Average Accel., km/h/s	2.25	2.30	2.37	2.41
Time, seconds	3.6	4.0	4.3	4.5
Distance traveled, m	45.0	65.0	90.0	110.0
Occupation of left lane:				
Time, seconds	9.3	10.0	10.7	11.3
Distance traveled, m	145.0	195.0	250.0	315.0
Clearance Length, m	30.0	55.0	70.0	90.0
Opposing vehicle, distance traveled, m	95.0	130.0	165.0	210.0

EXAMPLE PROBLEM 1.1

A tangent of bearing N31°30'W intersects another of bearing N40°10'W. A horizontal curve of 800 meters radius is designed. The design speed of the two-lane highway in desert terrain is 70 km/h. The Point of Intersection (PI) is Station 10 + 85.

1. What is the maximum grade?
   From Table 1.3, 7%

2. What is the central angle of the curve?
   40°10' − 31°30' = 8°40'

3. What is the PC Station?
   Tangent distance = 800 tan ½ 8°40' = 800 × 0.0758 = 60.5 m
   Sta. 10 + 85 − 60.5 = Sta. 10 + 24.5 = PC

4. What is the PT Station?
   Arc distance = Δ/360° πd = 8.667/360 × 3.1416 × 1600 m = 121.014m
   Sta. 10 + 24.5 + 121.0 = Sta. 11 + 45.5 at the PT

5. *What is the superelevation?*
   Rural highway without snow and ice; 0.12

6. *What is the first station on the curve?*
   With stations at 50 m intervals, Sta. 10 + 50

7. *What is the deflection angle from the PC to the first station on the curve?*
   ½ × 360° × 25.5/1600 × ⅓ × 1416 = 0.913° = 0°54'47" at Sta. 10 + 50

8. *What is the chord distance from the PC to the first station on the curve?*
   Sta. 10 + 50 − Sta. 10 + 24.5 = 25.5 m

9. *What is the deflection angle to the last station on the curve before the PT?*
   ½ × 360° × 75.5/1600 × ⅓ × 1416 = 2.709° = 2°42', at Sta. 11 + 00

10. *What is the superelevation runoff distance? From Table 1.2, 0.12 superelevation.*
    From Table 1.1, 80 m runoff

CHAPTER 2
EARTHWORK

THE EARTHWORK PROFILE AND MASS DIAGRAM

With the preliminary horizontal alignment staked in the field, a vertical traverse run on the centerline, and a preliminary vertical alignment chosen, the next step is to conduct an earthwork study. The study will determine the fine-tuning of the horizontal and vertical alignments. It produces an earthwork profile that compares the existing ground elevations and planned final elevations at each station, a tabulation of cut and fill volumes between stations, and a mass diagram that shows earthwork quantities that will be accumulated from cuts and used as fill throughout the project. Contract bid items, including excavation, haul distances and quantities, and borrow and waste quantities can be calculated using these data. Balance Points are stations on the mass diagram at which the quantities of excavation and fill have come into balance. Plotting the balance points on the earthwork profile will facilitate calculation of haul distances. (See Example Problem 2.1).

The first step in acquiring earthwork data is to level a traverse along the centerline and to plot the earthwork profile between the existing ground and the planned profile. The traverse is closed by making proportionate adjustments that distribute the corrections of error. The second step in the field work is to determine the limits of earthwork at each station in order to plot the cross-sectional areas between the existing ground and the final roadway, and calculate the amounts of cut and fill between each pair of stations.

CROSS SECTIONS: CUTS AND FILLS

At each station of the horizontal alignment, possibly excluding the beginning and ending stations of curves, the cross section must show an area that is perpendicular to the horizontal alignment and enclosed by the existing ground levels and by the proposed levels after roadbed construction. Cut and fill sections are measured in square meters. It is important to note the location of transition stations. A transition can be a station of near 0 cut and fill or a station where cut and fill are approximately equal. According to the commonly used Average End Area method, the volume of excavation or fill in cubic meters between two stations is the total area shown on the cross sections of the two stations, divided by two, times the distance between the two stations.

$$V \text{ (in cubic meters)} = A1 + A2 \text{ (square meters)} \times \tfrac{1}{2} \times \text{Distance (meters)}$$

The basis of the cross section is the design roadway from outer shoulder to outer shoulder. In difficult terrain, shoulder, lane, or median widths can be narrowed, fill or cut slopes can be made steeper, or retaining walls or stepped cut slopes can be used. These decisions can be made in the preliminary planning, or can result from the mass diagram data, or from difficulties encountered in staking the roadway widths.

On a cut section, field staking of the continuation from the shoulder line involves the inslope to the ditch and then the backslope to an intersection with the natural ground. The backslope can vary, depending on the material or the height above the ditch line. With a loose rock slope, the shoulder can be widened as a safety measure, and a fence or wall can be used to contain falling rock. On a very high cut slope, stepped cut slopes can control slides. Clay layers revealed during exploratory drilling or during construction can expand and slide upon exposure to surface rainfall. These clay conditions, or sand with a flat angle of repose, might make a retaining wall the most economical solution. (See Chapter 13 on Retaining Walls.)

On a fill section, the continuation from the shoulder can be a widened shoulder, especially for deep fills requiring a guard rail or edge drains. Next is the fill slope, which can vary depending upon the material used and the depth of the fill, to an intersection with the natural ground at the toe of the slope. For both cuts and fills, a reference stake is placed outside the work area, usually 5 meters from where the slope and natural ground intersect. The minimal volume of cut or fill that results because the slope rounds at the ground intersection can be disregarded.

With the elevation and distance from the center line of the outer limits of earthwork determined in the field and plotted on the cross sections, the volume of cut and fill can be calculated between each pair of stations. For a cut or fill that appears on one station but not on the adjacent station, an estimate of the 0 cut or fill station must be made to determine the portion of the distance between the stations to use for the multiplier in the formula. One of the cross sectional areas is 0.

WASTE AND BORROW

The mass diagram is a running measure of the amount of material excavated and available for fills, less the amount of material that has been used to build the fills. In rock excavation, a swell multiplier is used to calculate a reduction of fill material required. In all other excavation, an estimate of material lost (shrink) is used. The usual range of shrink is from 0 percent to 15 percent. It is applied to fill because of the need for replacement of the loss.

Construction is most economical when excavation and fill are close enough to be in the push range of bulldozers, so this is the first thing most earthwork contract bidders want to know. Next, they need to know the haul in the range of dump trucks used with front-end loaders. Beyond that range, excess excavation becomes waste, and excess fill requires borrow. When borrow is difficult to obtain or costly, longer hauls of material that would otherwise be waste are more economical.

Waste is seldom a problem. In fact, waste can improve comfort and safety when it is used to lengthen sag vertical curves or widen shoulders on fills. A large amount of waste material might even justify raising the fill grades. Shoulder widening is always useful, especially when it protects shoulder edge drains or provides more shying distance from the guard rail.

Borrow, on the other hand, can be very expensive and might justify major design changes. Some of these changes, such as narrower lanes or medians, or greater grade changes on sag vertical curves on fills, would reduce the highway design speed. Other design changes, which are preferred if they are economically feasible, would increase highway safety. In cuts, lower grade intersections and longer crest vertical curves, and wider shoulders especially on the inside of horizontal curves, would improve design and make more material available. If it is within haul distance, the added material would reduce borrow.

HAUL

Freehaul is the term for material moved a specified distance from the excavation to a fill, and is included in a contract as incidental to the excavation quantities. It includes the material pushed by a bulldozer and not actually lifted and hauled. Overhaul is material that is transported beyond freehaul distance but within a specified maximum distance. Overhaul is a bid item in contracts and the maximum distance influences the contract cost, but an increase in the maximum haul distance might be preferable to borrow. Design that minimizes both haul distance and borrow where possible reduces project cost.

Backhaul is haul from a station to a station that precedes it in the project. Ideally, the haul and backhaul from adjacent cuts would accomplish each fill with no borrow and little waste and would be within freehaul distance. Few situations are ideal, and designers usually compromise first on waste, next on overhaul, then on design augmentation or borrow, and finally on design reduction only if they must.

EXAMPLE PROBLEM 2.1

A project begins at Station 0 at existing ground level and at a 0% grade. At Station 0 + 50, the cut at centerline is 2.2 m. The roadway width from centerline to shoulder is 4.5 m. The ditch depth is 0.5 m, the ditch inslope is 4:1, the ditch bottom is 1.0 m, and the backslope is 2:1. The grade of natural ground perpendicular to the centerline at Station 0 + 50 is 2%. The earth excavation has a 10% shrink factor. The Freehaul distance is 200 m and the Overhaul distance is 400 m.

ADDITIONAL DATA

Sta.	1+00	1+25	1+50	2+00	2+50	2+90	3+00	3+50	4+00	4+50	5+00
Cut/Fill m	C1.1	C0	F1.6	F2.7	F1.1	C0	C0.6	C2.3	C0	F1.1	F0.6
Cross Sect. Area (sq. m)	20	0	25	40	15	0	10	40	0	20	10

1. *Sketch and label the cross-section at Station 0 + 50.*

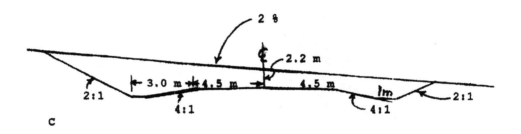

2. *Calculate the excavation area at Station 0 + 50.*
   Ditches: $2 \times 3m \times 0.5m \times 1/2 + 1m \times 0.5m \times 2 = 2.50 \, m^2$
   Rdwy to centerline elev.: $4.5m \times 4.5m/40 \times ½ \times 2 = 0.51$
   Assuming 1:40 crown
   Ditches to centerline elevation: $4m \times 0.1125m \times 2 = 0.89$
   Road and Ditches above centerline elev: $2m \times (9 + 8 + 0.225)m = 34.45$
   Cut slope angles above centerline: $2 \times 2.2m \times 4.4m \times ½ = 9.68$
   Total: $48.03 \, m^2$

   Short cuts: The area above the centerline is treated as a rectangle since + and − triangles are equal. The average cut slope triangle is doubled, rather than calculating each.

3. *Calculate the volumes of excavation and fill between Stations 0 and 5 + 50, and the accumulated fill material available at each station.*

4. *Draw a Mass Diagram from Station 0 to Station 5 + 00, at any convenient scale.*

EARTHWORK 2.5

To Sta.	Square Meters	Average	Distance	Exc.	Cubic Meters Fill & Shrink	Material
0+50	0 + 48	24	50	1200		1200
1+00	48 + 20	34	50	1700		2900
1+25	20 + 0	10	25	250		3150
1+50	0 + 25	12.5	25		312.5 +31.2	2806.3
2+00	25 + 40	32.5	50		1625 +163.5	1018.8
2+50	40 + 15	27.5	50		1375 +137.5	-493.7
2+90	15 + 0	7.5	40		300 +30	-823.7
3+00	0 + 10	5	10	50	-773.7	
3+50	10 + 40	25	50	1250	476.3	
4+00	40 + 0	20	0	1000	1476.3	
4+50	0 + 20	10	50		500	926.3
					+50	
5+00	20 + 10	15	50		750+75	101.3
TOTALS				5450	4862.5 +486.2	

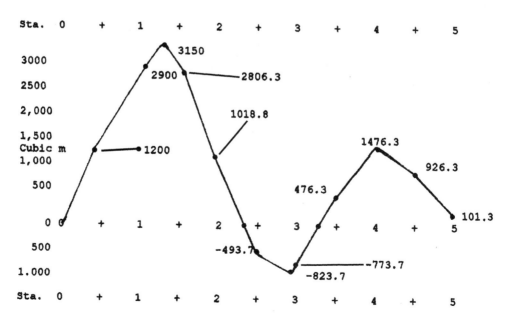

5. Draw a Profile Diagram showing the existing ground line and proposed grade from Station 0 to Station 5 + 00, with the same horizontal scale as the Mass Diagram.

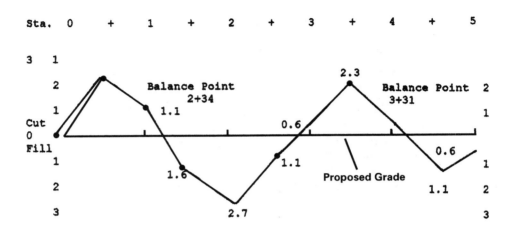

6. List the Stations at the volume balance points.
Answer: Station 2 + 34 and Station 3 + 31.
   $1019/1512 \, (50) = 34$
   $774/1250 \, (50) = 51$

7. What is the amount of Overhaul in this section of the project?
   By observation, haul cannot exceed 200 meters.
Answer: 0

8. What is the amount of Waste in this section?
   All excavation is used in an adjacent fill, assuming there is a fill beyond Sta. 5 + 00.
Answer: 0

9. Calculate the total haul from Station 0 to Station 5 + 00, in cubic meters hauled 100 meters.
   $3150 \times (0 + 234)/2 \times 1/100 = 3685.5$
   $623.7 \times (331 - 234) / 200 = 302.5$
   $(1476.3 - 103.3) \times (500 - 331) /200 = 1160.2$
Total: 5148.2

10. What is the total shrinkage adjustment?
Answer: $31.2 + 162.5 + 137.5 + 30 + 50 + 75 = 486.2 \text{ m}^3$

CHAPTER 3
DRAINAGE

On most streets and roads, the cost of drainage is a major item in construction projects. The cost of inadequate drainage, however, can be far greater. Proper drainage is important to the safety and convenience of road users, prevention of damage to adjacent property, control of environmental damage, life of the construction, and the cost of maintenance.

CURBS AND GUTTERS

Curb and gutters carry runoff water from an urban street to a storm sewer or drain channel, usually located at the next intersection. The minimum grade of the street needed for this drainage is 0.5 percent. Heavy runoff or substantial distance to a storm sewer or other relief point might cause a problem when a minimum grade is provided.

Gutter construction allows moving traffic or parked cars to use roadway width that would be the shoulder in rural areas. Curbs protect pedestrians from traffic and protect traffic from street-side obstacles. Where a barrier curb is not needed for auto and pedestrian safety, a mountable curb can be installed. A mountable curb would have a maximum height of 150 mm and a maximum slope of 1:1, with desirable heights and slopes much less.

The minimum curb height above gutter level should be 125 mm, which can be reduced to 100 mm in special circumstances. Depending upon drainage and safety needs, heights can increase to a maximum of 225 mm. If a higher curb is essential, it should be sloped rather than vertical, or it should be offset at least 0.3 m. In combination curb and gutter construction, the gutter pan normally provides an offset of 0.6 m.

Curb cut facilities for wheelchair users are much more economical when planned in original construction rather than built as retrofitting on existing streets. Lane width for wheelchairs should be no less than 1.0 m, with safety islands no less than 1.2 m and 1.8 m desirable where wheelchair turning is required.

Barrier curbs are of portland cement concrete, or of granite when resistance to road salt makes granite more economical. Mountable curbs can be of portland cement concrete or asphalt concrete and can be poured or extruded. Barrier curbs and mountable curbs generally should not be used in rural areas, since they can become hazards in darkness or during snowfall.

DITCHES AND EDGE DRAINS

Roads are crowned to drain water from the roadway surface to the fill shoulder or to the ditch. The ditch, in addition to carrying away the surface runoff, drains the subgrade of the road to prevent damage to the surfacing. Water under the pavement can liquefy the subgrade and cause a pothole, or it can break up the pavement in a freeze and thaw cycle.

The accumulated water in the ditch is conveyed to a natural stream for discharge. The minimum ditch grade is 0.5 percent, but a greater slope should be provided if heavy runoff is anticipated or if the ditch carries water a considerable distance to a relief point. The gradient and alignment of the ditch in most cases should be the same as that of the road, but they can be varied as necessary to achieve the best design.

The objective in ditch design is to protect the road from damage and to protect road users from danger and inconvenience while making minimal changes in natural drainage patterns. Depending upon the roadway material and the grade, partial dams might be needed to slow the water flow and prevent ditch erosion. Erosion seldom occurs in rocky material, or in any material at a grade of 1 percent or less. Ditches on the downhill side of the road can be built to discharge water into natural drains or onto grasslands where the water can spread without causing erosion.

Open channel flow, as in ditches, is turbulent and thus more difficult to measure than flow in a pressure conduit flowing full. A ditch designed to drain subsurface water from under pavement usually will be large enough to carry runoff water to the nearest natural channel. Ditch depth should be at least 0.5 m, with a preferred depth of 0.6 m or more. The ditch must be widened when it collects substantial flows before the water can be discharged to a natural stream.

Open channels are more efficient if slopes are as steep as the material will stand, but those adjacent to a roadway must have rounded, flat slopes to help drivers who leave the road regain control of their vehicles. For minimal construction, ditches can have 4:1 inslopes, or even steeper on roads with very little traffic. To provide recovery opportunity for errant vehicles, however, 5:1 or flatter slopes are recommended, along with shoulder (hinge point) rounding and ditch-bottom rounding. Shoulder rounding has been found to reduce the tendency of vehicles to go airborne, and ditch-bottom rounding avoids trapping the front wheel and allows directional recovery. In northern regions, 6:1 and flatter slopes are recommended to reduce snowdrifts, and flatter slopes make maintenance mowing easier.

Ditches and channels should be lined only to the extent necessary to control erosion. A channel will not necessarily follow the grade or parallel the alignment of the roadway.

Seeding and mulching of natural ground cover on fill slopes, ditches, and drainage channels are desirable to control erosion, reduce maintenance, and beautify the roadway. Maintenance mowing is facilitated by flat ditch inslopes.

If a sag vertical curve with plus and minus grades is in a cut, the point of discharge must be near the grade intersection. Care is required in design to avoid water pooling in the ditch. A similar situation might exist in crest vertical curves, where a very flat curve has a rate of curvature (length of curve divided by the percent of change of grade) exceeding 51 m. On these curves, the summit might have an inadequate grade for too long a distance to provide proper drainage along the ditch.

Vegetation usually covers fill slopes. Water flows over the shoulder and down the slope causing little erosion. When this is not the case, a shoulder drain, usually molded of asphalt at the shoulder point, is designed to carry the water to a location where it can combine with water in a ditch or culvert and be controlled properly.

CULVERTS

Ditches on the uphill side sometimes discharge into a natural watercourse, but on long slopes they usually are discharged by a dam built across the ditch that diverts the water into a culvert under the road and into the ditch on the other side or into the natural drainage. The cross culvert should be perpendicular to the road centerline to reduce cost, unless there is a design reason to install it at an angle.

Culvert size is determined by the expected volume and rate of rainfall and the size of the runoff area served by the culvert, along with the grade of the culvert. A cross culvert that will drain the uphill ditch usually is designed for a 25-year storm. A cross culvert in a defined stream crossing a minor highway usually is sized for a 50-year storm. A cross culvert in a defined stream crossing a major highway usually is designed for a 100-year storm.

If a culvert across a road or in a downhill ditch must discharge onto a fill, slope erosion must be prevented. This can be accomplished by continuing the culvert, or installing a half-culvert or a channel protected by rock or vegetation, to the toe of the slope. An installation of concrete or stone at the toe of the slope should absorb the energy of the rapidly moving water. Culverts also are used in or near a ditch and parallel to the ditch flow line to allow water to pass under an intersecting road or driveway.

Concrete pipe is a permanent installation, unless it is damaged by an earthslide or earthquake or undermined by erosion. Many installations have less expensive alternatives. Corrugated steel pipe has been available since 1896, and it can be specified in full round diameters of 6 in. to 144 in. (0.15 m to 3.66 m) and in thicknesses from 0.16 cm to 0.43 cm. Helical corrugated pipe is also available, and it is stronger and more watertight because the seam is spiraled.

Large-diameter, high-density polyethylene pipe has been used for natural gas transmission for years, and more recently it has been approved for highway drainage up to 48-in. (1.22 m) diameter by the American Association of State Highway and Transportation Officials (AASHTO) and by the Departments of Transportation in all 50 states. Transportation officials have described the pipe as "durable, well-performing, and cost-effective."

Pipe-arch pipe, with high capacity in low-headroom conditions, also can be specified. Structural plate pipe is available in larger sizes, with diameters from 1.5 m to 8 m. These pipes are of heavier gauge and are made of plates that are bolted together on-site. They are available as full-round, elliptical, arch, or underpass shapes, and are used for drainage, streams, small bridges, and pedestrian or vehicle underpasses.

Heights of cover to 30 m are permissible. The zinc coating is adequate under most conditions. For extremely corrosive conditions, a hot-dip coating can be specified. If very corrosive conditions are expected, a paved invert can be installed to protect the invert of the pipe.

The most common method for determining runoff for transportation structures, excluding major bridges, is the *rational* formula, which was designed to avoid excessive calculations for minor changes:

$$Q = CIA$$

where Q = Peak discharge in cubic centimeters per second
C = Runoff coefficient, the percentage of rain that is direct runoff
I = Rainfall intensity, in cm per hour
A = Drainage area, in hectares

PRECAST CULVERTS AND SMALL BRIDGES

Precast concrete culverts might be required when a corrugated culvert is not large enough to carry the anticipated flow, when the top of the corrugated culvert is too near the finish grade of the road, or when erosion might undermine the culvert. Precast portland cement concrete box culverts are available in many sizes and configurations or can be cast to order at additional cost. Where necessary, more than one precast box culvert can be installed side by side in the same stream. An alternative is the reinforced concrete bridge culvert, a box that is open on the bottom to allow the stream to flow naturally.

Bridges longer than 30 m or in problem locations require individual study, analysis, and design. (See the PE series book titled *Structural Engineering*.) Bridges shorter than 30 m must have a design loading structural capacity of AASHTO's MS-18 rating. Small bridges must be no less than full roadway width if they are on rural or urban arterials, or on collector roads and streets that are paved the full width from shoulder to shoulder or are more than 2000 Average Daily Traffic (ADT). For those collector roads and streets that are less than 2000 ADT and lack shoulder paving, the minimum bridge width requirement is the width of the traveled way plus:

m on each side	ADT
0.6	less than 400
1.0	400 to 1500
1.2	1500 to 2000

Bridges can remain in place rather than be replaced on reconstruction projects for secondary roads and streets if they meet the following requirements:

MS-13.5 Design Loading Structural Capacity

Fits the proposed horizontal and vertical alignment

Has minimum clear width between rails and curbs of 6.6m for ADT less than 1500, 7.2m between 1500 and 2000 ADT, and 8.4m higher than 2000 ADT.

CHAPTER 4
PROTECTION OF THE ENVIRONMENT

AVOIDANCE OF ENVIRONMENTAL DAMAGE

Highways and railroads are a part of our environment. It is essential that designers consider them so, and strive for maximum positive impact and minimum negative impact on other parts of the environment. Routes that minimize damage should be chosen, unless cost is prohibitive or proper road user service cannot be achieved. Construction should be planned to avoid creating problems that might be objectionable to adjacent land users and to alleviate existing environmental problems if possible within the parameters of the road project.

NOISE POLLUTION

Unless a road is in an unpopulated area, there will be noise pollution. With sparse traffic or a sparse population, the problem might be minor and require no mitigation. Noise is seldom a problem to road users in cars designed to minimize road and wind noise, but easily might become a problem to residents near a road. Most noise originates from tire contact with the road, and the sound travels in a straight line until it reaches a substantial obstruction. To be substantial, the obstruction should be at least a solid $^3/_4$-in. wooden fence, but it can be a rock or concrete wall or an earth berm. The noise barrier must be high enough to shield the ears of adjacent residents from the line between the tire-road contact point and the top of the fence, extended.

The major exception is noise from truck exhaust, which is greatest when trucks are negotiating a long, heavy grade. Because it is difficult to make a noise barrier high enough to shield the high truck exhaust, constructing such grades in residential areas should be avoided unless the road is in a substantial cut. When these criteria cannot be met, the possibility of noise pollution damage to adjacent properties should be factored into economic analyses.

Noise barriers on the inside of curves should be placed carefully so they do not obstruct the line of sight for safe stopping distance. Barriers should not be placed within 60 m of the gore areas of interchange exits.

HISTORICAL SITES, PUBLIC BUILDINGS, AND PARKS

Careful route selection and design is best to avoid environmental damage by road construction to historical sites, public buildings, parklands, and other recreational land. When environmental damage cannot be avoided entirely, it should be held to a minimum and mitigation actions should be planned. The reasons why the damage is unavoidable and the mitigation actions that will have to be taken must be explained in public meetings, along with route decisions and road designs.

A construction site that contains pre-Columbian Native American artifacts or articles used in early European settlements will require archeological investigation. Based on that investigation, an archeological dig may be required to protect and preserve such objects, or a rerouting of the planned construction might be necessary to avoid the site.

WETLANDS FILLING

Of all potential damage sites, the most unavoidable because of their size, and yet most amenable to mitigation and replacement are wetlands. When a road crosses wetlands, it should fill only the area needed for road operation to minimize the wetlands reduction. While acquiring borrow material for the road fill, new replacement wetlands can be created near the roadway. Replacement should assure that more wetlands exist at the completion of the project than existed at the beginning.

After borrow excavation, the most efficient way to replace wetlands is to buy drained farmland and to return it to wetland status. Most of our wetlands loss has been agricultural, and a high percentage of this loss preceded legal controls on wetlands conversion. The narrow ribbon of land for a road is a small area compared to farmlands. A search for the least productive drained farmlands would reduce the cost of purchase, identify the most likely sellers, and minimize the reduction of agricultural produce. Mitigation by replacement reverses the loss of wetlands that, as we now know, are essential for water purification, as well as for wildfowl habitat and erosion reduction.

New and recovered wetlands should be designed carefully, primarily with native plants and with water supply as near as possible to that of the original wetlands. Topography should be in harmony with neighboring properties, especially other wetlands.

EROSION AND SILTATION

A roadway acts as a long obstruction, seldom parallel to surface water flow, with a surface that must be kept as dry as practicable. The road construction, therefore, will change preexisting drainage patterns. Vegetation removal and surfacing, especially paving, will accelerate runoff. This unavoidable environmental damage must be minimized in the design.

A design that protects the road from unusually large rainfall events also will reduce the occurrence of siltation damage to the environment. Consideration of this factor might make the use of riprap or geotextile slope protection feasible.

Damage during construction is usually due to erosion or siltation and must be controlled by temporary facilities such as plastic reinforced dams or straw bales that will confine silt to the work site.

Erosion and siltation can be minimized by seeding new fill and cut slopes, preferably with native grasses. Fill slopes should be left rough and cut slopes serrated to permit percolation of rainfall and deter runoff. A blend of seed, fertilizer, and mulch can be sprayed on the slopes, and watered between rains until root systems are established.

Grass is the most economical erosion control on steep slopes with little water flow and flat slopes with heavy flows. Channels can be lined with concrete, asphalt, rock, or other materials when the velocity is such that grass cannot control the erosion. Very high velocities can be controlled with pipes leading to energy-dissipating structures of concrete or rock.

Tree plantings can provide a root system for permanent control of erosion and create beauty for the pleasure of the road user. They should be planted where they will not be a roadside hazard or an obstruction to sight distance on the inside of curves. Varieties consistent with the local climate should be used. Trees also might be preferable to fencing to screen junkyards and other such sights from view from the road. They should be considered as a screen at the right-of-way line, between the road and residential areas either through retention of natural growth or new plantings. To be an effective noise barrier, planting areas must be a least 3m wide and densely planted.

Trees, boulders, and other unyielding objects should be removed from a clear zone adjacent to the roadway. The clear zone should be a minimum of 3.0 m wide for low-volume, low-speed roads, and much more for heavily traveled roads. For high-speed freeways, the clear zone should include all areas an errant car is likely to reach, except those well below the roadway.

Permanent structures especially should be designed to minimize damage caused by the presence of the road. Redirected and increased runoff should be routed to streams capable of handling the increased volume. Check dams of stone, asphalt, or concrete will reduce erosion in ditches by slowing the rate of water flow. Siltation from unavoidable erosion should be held on or near the roadway.

Final stage siltation control can be completed with a dam at the downstream margin of the right of way, with an overflow channel and a sediment basin. Sometimes a culvert is embedded in the dam lower than the overflow channel to begin stream flow earlier and continue it longer, and to protect the overflow channel from excessive flow. The dam will decelerate flow and release it at a rate consistent with the capacity of the natural channel, and collect sediment in a location where maintenance forces can clean it out and use it for fill.

AIR POLLUTION

In dry weather, water spraying will be necessary to control dust blowing off a construction work site, especially in populated areas. This also makes the site more comfortable for the workers and assists in compaction of earthwork. On earth and gravel roads in populated areas, regular maintenance spraying with a chemical solution might be necessary to control dust and to retain fine particles in the road surface.

Fumes from asphalt construction must be kept within reasonable concentrations. While seldom a major problem in transient paving operations, fumes might be a major factor in the siting of an asphalt plant. It might be necessary to locate the plant in an unpopulated or sparsely populated area, still within a convenient haul distance from the construction project.

Construction inspectors should be aware of their responsibility for preserving air quality, and insist on proper maintenance of the engines of construction machinery.

WILDLIFE HABITAT

Inevitably, wildlife habitat will be taken in any road construction. At the beginning of the food chain, wildlife exists in teeming millions in every sample of topsoil. For the higher animals in the food chain, the pavement becomes a desert, as well as a danger. The taking of habitat and food source can be minimal and of little consequence in relation to the size of the habitat available, or a substantial portion of the available habitat could be destroyed and be a major concern.

If endangered or threatened species in limited habitat are involved, alternatives to the damage must be studied. The alternatives are avoidance, relocation, and habitat replacement. Serious consideration must be given to avoiding the habitat by rerouting the project, because the alternatives of relocation of species and habitat replacement are problematic and likely to be expensive both in time and money.

An environmental assessment must be made to evaluate and enumerate the environmental effects of the roadway construction and include the major species that might be affected. If investigations discover major environmental impacts, especially on threatened species, a careful and complete Environmental Impact Statement must be prepared, in cooperation with local government and with public notice.

EXAMPLE PROBLEM 4.1

1. What damage is always present to some degree in road construction, and must be planned for mitigation?

    A. Taking of historical sites.

    B. Taking of recreational areas.

    C. Changing surface water flow patterns.

    D. Moving earth materials.

Answer: C

2. What is usually the first factor to consider in the mitigation of wetlands filling?

    A. Locating drained farmland.

    B. Locating sources of borrow.

    C. Planning use of native grasses.

    D. Cost of land for replacement of filled wetlands.

Answer: B

3. *Most wetland loss has been due to:*
   A. Road construction.
   B. Railroad construction.
   C. Building construction.
   D. Farmland drainage.

Answer: D

4. *Road construction accelerates runoff from rains because of:*
   A. Interfering with natural water flows.
   B. Surfacing and paving.
   C. Ditch construction.
   D. Culverts and bridges.

Answer: B

5. *An environmental damage that should nearly always be avoided rather than mitigated is:*
   A. Taking of historical sites.
   B. Filling in wetlands.
   C. Changing patterns of surface runoff flow.
   D. Erosion.

Answer: A

6. *Public hearings must be held for:*
   A. Route selection.
   B. Environmental damage avoidance and mitigation.
   C. Road designs.
   D. All of these.

Answer: D

7. *The wetlands taken are the area:*
   A. Within the road right of way.
   B. Under the fill.
   C. Between the shoulder lines.
   D. Between the ditch lines.

Answer: B

8. *Permanent construction should prevent:*
   A. Siltation of adjacent property.
   B. Increase of water flow in culverts under the roadway.
   C. Increase of surface water runoff rates.
   D. Increased water volumes in natural streams.

Answer: A

9. *Temporary construction dams are intended primarily to:*
    A. Direct water flowing off the work site.
    B. Contain surface water.
    C. Contain siltation from work-site erosion.
    D. Prevent erosion.

Answer: C

10. *Many wetland replacements probably will be from:*
    A. New excavation and construction.
    B. Raising the water level.
    C. Returning drained farmland to wetland status.
    D. Building dams to create wetlands.

Answer: C

CHAPTER 5
ROADWAY DESIGN

DESIGN VEHICLES AND TURNING WIDTHS

In roadway design, consideration is given to two-wheeled vehicles (bicycles and motorcycles), passenger cars, motor homes with or without trailers, buses (single-unit or articulated), and trucks (single-unit to turnpike double semitrailer). The control vehicle for horizontal curve design is usually the largest truck permitted on the road. Design review, however, should assure the curve's compatibility with other types of permitted vehicles. On most high-speed highways, for example, bicycles are permitted only on specially designed shoulders or are not permitted at all.

The American Association of State Highway and Transportation Officials (AASHTO) has determined vehicles' turning characteristics that are critical to roadway design: the minimum design turning radius and the minimum inside radius. The minimum design turning radius is the path of the outside front overhang.

The Surface Transportation Assistance Act of 1982 (STAA) adopted a design vehicle with a 13.7-m turning radius, which is determined by the path of the outside front overhang. This 13.7-m turning radius is the same for a large semitrailer combination (SB-15), a semitrailer-full trailer combination (WB-18), and an interstate semitrailer (WB-19, with a 14.6-m trailer, or a WB-20, with a 16.2-m trailer). The WB-19 was adopted as the design vehicle in STAA, and the WB-20 was grandfathered in as the design vehicle in some areas. These are the usual control vehicles for horizontal curve design.

Triple semitrailer trucks (WB-29) and turnpike double semitrailers (WB-35) are permitted on a few highways in some western states. They each have a minimum turning radius of 15.2 m and 18.3 m, respectively. A motor home with a trailer can have a minimum turning radius of 15.2 m. (See Table 5.1).

TABLE 5.1 Design Vehicle Minimum Turning Radii

	Effective Wheelbase m	Min. Design Turn Radius m	Min. Design Inside Rad. m	Pvt Min. Width m
Passenger Car	3.4	7.3	4.2	3.1
Car with Trailer	4.6	7.3	2.0	4.5
Motor Home	6.1	12.2	7.9	4.3
MH with Trailer	4.6	15.2	10.7	4.5
Single Unit Bus	7.6	12.8	7.4	5.4
Articulated Bus	5.5	11.6	4.3	7.3
Single Unit Truck	6.1	12.8	9.5	3.3
Small Semitrailer Truck WB-12	8.2	12.2	5.7	7.5
Large semitrailer Truck WB-15	9.1	13.7	6.8	6.9
Large Semitrailer Truck WB-18	6.4	13.7	6.8	6.9
Large Semitrailer Truck WB-19	12.8	13.7	2.8	0.9
Large Semitrailer Truck WB-20	14.3	13.7	0	13.7
Triple Semitrailer WB-27	6.6	15.2	6.3	8.0
Turnpike Double Semitrailer WB-35	13.4	18.3	5.2	13.1

The path of the inside rear wheel defines each design vehicle's minimum inside radius. On a freeway or expressway with a 3.6-m lane width, the design semitrailer will stay in the lane if the curve of the inside pavement edge has a radius of at least 23 m for the small semitrailer (WB-12). That minimum radius is 29 m for the WB-15 and 35 m for the WB-18 (usually the design vehicle of choice). Lane widths that preserve shy distances for adjacent traffic lanes and that promote smooth traffic flow would require longer radius curves. The tightest curves, suitable for traffic that is traveling less than 15 km/hr, usually are used when a small amount of space is available—usually in unchanneled intersections. When traffic is moving at higher speeds, flatter curves would be required and lane widening would be less extreme.

On four-lane streets in urban areas, all vehicles can make left turns with little encroachment on opposite lanes. For right turns, a curve radius greater than 4.5 m (preferably 7.5 m to 9 m), is adequate for passenger cars. A radius of 12 m or larger is required for trucks and buses. Trucks above WB-15 will encroach on the third lane while turning, with WB-35 taking up the entire third lane.

LANES

A lane width of 3.6 m is required in freeway construction and is strongly recommended elsewhere. In urban settings with expensive right-of-way or in low traffic conditions, 3.3 m is a valid compromise. For low-volume rural roads, a 3-m or even 2.7-m lane width is acceptable (Table 5.2). Lane width in curves, as has been shown, is a function of the design truck's turning requirements. On tangents, the standard lane width accommodates truck width plus shying distance from vehicles in adjacent lanes. On curves with long radii, the extra width requirement of turning trucks is negligible.

TABLE 5.2 Percent Capacity Loss: Narrow Lanes with Narrow Shoulder or Obstruction

	Lane Width			
	3.6 m	3.3 m	3.0 m	2.7 m
Clearance Four Lanes				
1.8 m	0	5	11	23
1.2 m	2	6	12	24
0.6 m	5	8	14	25
0	12	15	20	30
Clearance Two Lanes				
1.8	0	7	16	30
1.2	8	15	23	35
0.6	19	25	32	43
0	30	35	42	51

As curves become sharper, lanes should become wider to safely accommodate tracking differences between a vehicle's front and rear wheels. A truck that is 2.6 m wide has 0.5-m shying distance on each side of a 3.6-m lane. If the tracking difference is more than 0.2 m, lane widening should be considered, especially on high-traffic roads. If the difference is more than 0.4 m, lanes should be widened or other safety measures should be taken, such as a reduced speed limit or warning signs. In highway design with superelevated curves and design speeds greater than 60 km/h, track width differences for trucks are not significant except for WB-35. WB-35 would require lane widening below a radius of 400 m; WB-20, 200 m; WB-19, 180 m; WB-29, 160 m; and WB-15, 100 m.

Lane cross slopes on high-type pavement should be at a 2-percent grade. Urban streets can be 1 percent or 1.5 percent. When the pavement could have rutting or differential settlement, slope should be 3 percent or more to promote drainage. Gravel or earth roads should slope 3 percent to 6 percent. Slopes less than 3 percent can be used in climates where heavy rainfall is unlikely. A rounded crown should be used where traffic crosses the centerline while passing.

SHOULDERS AND DITCHES

Shoulders on all but the lowest-volume roads should be wide enough for emergency parking. On all paved roads, especially those with heavy traffic, shoulders should be paved when construction economics permit. Paved shoulders are not only safer and more convenient for drivers, but they move runoff water farther from the pavement edge. This saves money on pavement maintenance and recoups the cost of the original paving.

Shoulder width on high-volume highways should be at least 3 m and preferably 3.6 m, especially when truck Directional Design Hour Volume (DDHV) exceeds 250. Shoulders of the lowest-volume highways should be at least 0.6 m and preferably 1.8 m to 2.4 m. Shoulder width should be continuous if possible, with extra widening for the guard rail, signage, and lighting. Shoulder width across bridges is an effective safety and operating measure, and it should be done except in extreme circumstances.

Freeway right shoulders should be paved. Median shoulders also should be paved, and they should be at least l.2 m wide when there are four lanes, 3 m wide when there are six or more lanes, and 3.6 m wide when truck traffic exceeds 250 DDHV. Shoulder slopes should drain water away from the pavement effectively, but they should not be so steep that motorists cannot use them in emergencies. Experience has shown that paved-shoulder slopes of 2 percent to 5 percent work best. The steepest slopes are used in areas of heavy rainfall.

For gravel shoulders, the best range is 4 percent to 6 percent, and 6 percent to 8 percent works best for grass. Shoulders should meet the pavement edge smoothly; this is an inspector's responsibility on paving projects and a regular maintenance task on unpaved shoulders. The shoulder at the top edge of a superelevation might need to be adjusted so that no grade change exceeds 10 percent. An additional grade change in the middle of the shoulder might be necessary to accomplish this.

Ditch inslopes should be no steeper than 4:1, preferably 6:1 to 10:1. The flatter inslopes provide a recovery area for straying vehicles, and they help to prevent many one-car accidents. The shoulder edge, called the hinge point, should be rounded to help prevent straying cars from becoming airborne. The ditch bottom should be widened and rounded because a V-shaped ditch could trap a vehicle's front wheels.

The backslope can be any convenient slope that fits the need and the terrain, from 10:1 to acquire material for a fill to 0.1:1 in a solid rock cut. A shallow backslope also might add to road safety, but this is seldom considered economical unless fill material is needed or passing sight distance on a curve is required.

MEDIANS

Medians are used on urban expressways to prevent head-on collisions (the greatest danger in high-speed traffic) and to provide space for left-turn channels. A minimal median would have a double stripe, curb, or traffic divider with minimal shying distance. If left turns are not prohibited, traffic in the left lane would be permitted to turn left or go ahead.

Ideally, an urban median would be 5.4 m wide in order to have a 3.6-m left turn lane and a 1.8-m curbed separator. If space is at a premium, a 3-m lane and a 0.6-m separator will work well. In restricted circumstances, a continuous left turn lane can be fitted into a 3-m space. Experience has shown that narrow medians work well, as do medians that are wide enough (above 15 m) to provide storage for two separate parts of the left turn.

Medians that are not obviously wide or narrow, between about 9 m and 15 m, have shown operating difficulty because of drivers' confusion about the median's purpose; drivers do not know if they should complete the left turn or pause until the other light turns green. When a median is narrower than 3 m on a road with substantial traffic, it might be best to prohibit left turns, forcing traffic to make three right turns on adjacent streets to accomplish the movement.

In urban areas with expensive right-of-way or in mountains with expensive earthwork construction, multi-lane highway medians can be as narrow as a double paint stripe or a median barrier, or so wide as to effectively make two independent roadways. Most often, two parallel roadways have at least enough space between them to facilitate drainage, and they are likely to contain the space needed for future lanes and to support other transportation modes.

Medians in rural freeways and in urban settings with adequate width provide excellent opportunities for drainage that will protect the roadways. Highways with wide medians and those of substantially different elevation can have separate roadway drainage systems in the median. On highways with narrower medians, this usually is a joint drainage system in a depressed median.

CHAPTER 6
TRAFFIC ESTIMATION AND DESIGN VOLUMES

TRAFFIC COUNTING AND ANNUAL INCREASE

Traffic counting has been a regular part of highway management and planning in the United States, Europe, Japan, and many other countries, for many years. Recording the annual traffic growth by highway, area, and region over many years can help predict future traffic volumes for proposed projects. Known and projected factors such as demographics, economics, and development can be used to modify the projected traffic trend and the traffic forecast.

The most recent traffic volume is the ideal beginning, since next year's volume probably will differ from this year's by a small amount. To look far into the future, however, we need to examine the past more closely. Have volumes been stable, increasing, or declining? If they have been increasing, which is usually the case, is the rate at which they are increasing accelerating or decelerating? What are the factors that cause acceleration? Are these factors stable, or are they changing? If the factors are expected to change, should an initial rate of increase be modified after a certain number of years? This study is likely to result in a forecast rate of traffic volume change that differs from the rate of the most recent past.

TRAFFIC FORECASTS

A horizon of at least 20 years is common for highway planning; at least 50 years is common for major bridges. These times, however, are not a rule, and the design year should be based on the circumstances of each project. For example, a 30-year highway design might have a reasonable additional cost, providing future public benefit that would justify the interest on the added cost. In another case, concern with the reliability of estimates or with other fiscal matters might limit the project to a 10-year design with the need for review within a decade.

The planning time horizon is a combination of the expected service life of the construction and the expected cost of modification or replacement. Major bridges are much more costly than highways and small bridges, and they are expected to be more durable. This gives them a bigger risk of premature obsolescence and might justify construction or additions that are not essential during the highway's planning horizon.

When a bridge has a 50-year planning horizon, it is advisable to estimate the 50-year traffic and lane requirements on the road as well, even though traffic estimation accuracy diminishes rapidly at more than 20 years. This will evaluate the bridge's possible obsolescence before its planned lifespan is completed. The dichotomy between bridge and highway life spans is the reason a bridge often is designed to serve both directions of traffic, with a parallel bridge and roadway to be built at a later date.

Traffic estimation is important. Because of political or economic problems, many highways on the Interstate System were built with fewer lanes than were needed according to the forecast. Most of these highways became congested and had to be reconstructed or relieved with parallel construction well before their design year.

Budgeting and project priority decisions usually are based on forecasts of Average Daily Traffic (ADT) for the design year. These forecasts are based on the total traffic projected for that year divided by the number of days. ADT is a good measure to compare projects. ADT usually is determined on an annual basis, but it can apply to other periods, such as weekend ADT, weekday ADT, summer months ADT, or any other division the planner desires, if the data to support the traffic estimate are available. If the ADT is only known for one direction, the total road ADT is usually twice the directional ADT.

LANE CAPACITY

ADT is an important basic traffic study, but a project designed and built for the ADT would be congested half the days in the design year. This degree of congestion is unsupportable in a design, but designing for no congestion would be expensive and would produce a road that seldom reached near-capacity, even in the design year. The Design Hour Volume (DHV) is the compromise measure used for road design.

DHV, usually defined as the 30th high hourly volume (30HV) forecast in the design year, accepts that 29 hours of congestion will occur in the design year. It is a reasonable compromise in nearly all cases. In unusual cases, such as a road serving a summer resort or a ski area, 10HV could be triple 30HV. This degree of congestion could be deemed unacceptable even a few hours a year, and the DHV for that project should be changed to 15HV.

DHV is calculated by measuring traffic volume on an hourly basis over a year, conducted for several years, and extended into a forecast for the design year. If this data is not available, as it would not be for a new highway location, data from a nearby comparable highway or the estimate DHV = $0.15 \times$ ADT can be used. Studies have shown that 30HV is between 0.12 and 0.18 of ADT on more than 70 percent of roads.

Traffic peak loads dominated by commuters tend to be heavy in one direction in the morning and heavy in the opposite direction in the evening. A typical directional traffic distribution at the peak hour is 60:40, but variations might be substantial. Business growth causes the peaks to become exaggerated, both in the morning and in the evening. Trends might develop so that the differences become less; businesses migrating to the suburbs would cause a reduction in the morning peak and might cause some reverse commuting.

If directional studies are made, historical peak flows and daily changes of peaks, trends, and forecasts of future trends would result in directional DHV, or DDHV, forecasts for the design year. The DDHV for multi-lane roads is ADT times the percentage 30HV is of the ADT, times the percentage of traffic moving in the principal direction in the design hour. In the common situation of 60:40 traffic, DHV = 0.15 ADT, DDHV = (0.15)(0.60) ADT = 9 % of ADT.

Maximum lane capacity in free flow in a freeway lane is dependent upon speed. At vehicle speeds of 110 km/h (68 m/h), maximum lane capacity is 1300 passenger cars per lane per hour. At higher speeds, capacity declines because spacing between cars increases as drivers opt for more stopping space in case of emergencies. The spacing increase is a greater factor in capacity than the speed increase (see Table 6.1).

TABLE 6.1 Maximum Capacity

Speed, km/h	Maximum Capacity Passenger Cars /hr/Lane
120	1300
110	1450
100	1600
95	1690
90	1760
85	1830
80	1900
75	1970
70	2040
65	2100
60	2150
55	2200

At lower speeds, drivers close the distances between cars, so capacity increases. Lane capacity is greatest at about 55 km/h (34 m/h), at which the maximum lane capacity reaches 2200 passenger cars per lane per hour. When maximum capacity is reached, traffic density can increase, but speed declines rapidly and volume decreases until it reaches near zero for stop-and-go traffic conditions.

The instability inherent in the close proximity between maximum capacity and gridlock and the road users' desire to keep speeds above 55 km/h usually leads to a capacity goal of 2000 passenger car equivalents per lane per hour. The capacity goal might not be reached; gridlock can occur at lower traffic volumes because of the many factors affecting free flow, such as obstructions near the roadway, narrow shoulders, restricted sight distance, lane width less than 3.6 m, and merging traffic.

On urban arterials, other factors that reduce maximum capacity are intersections, traffic lights, and turning movements. If traffic stops, estimates of maximum resumption of traffic flow range from 1500 to 1800 passenger cars per hour. In these figures, the percentage of trucks is critical; a truck is the equivalent of 1.5 to six passenger cars, depending upon the terrain. See Table 6.2 for passenger car equivalents of buses, trucks, and recreational vehicles.

TABLE 6.2 Passenger Car Equivalents

	Level	Rolling	Mountainous
Buses	1.5	3.0	6.0
Recreation Vehicles	1.2	2.0	4.0
Trucks	1.5	3.0	6.0

LANES NEEDED AT DESIGN YEAR

The DHV estimated for the highway's design year, divided by each lane's capacity, determines the number of lanes needed and possibly the phasing of construction. If the highway is to be a freeway with 2000 lane capacity, 20 percent of which is trucks and buses, and a DHV of 4500 in each direction, the directional passenger car equivalent will be 4500 + 4500 (0.2) = 5400. With 2.7 lanes needed in each direction in the design year, a decision might be made to build a six-lane freeway but delay the paving of the fifth and sixth lanes to a later date, possibly in the 10th or 12th year.

A four-lane urban expressway that has 3.3-m lanes, no trucks or interference from turning traffic, obstructions 1.2 m from the traveled way, and a signal that is green 50 percent of the time would have a maximum lane capacity of $1800 - 900 - (900 \times 0.15)$, or 765. (Tables 5.2 and 6.1).

This expressway would have a directional capacity on two of its lanes of 1530 per hour. If that capacity were exceeded, the choices would be to remove obstructions, add lanes, or redirect traffic. Moving the obstructions at least 2.5 m from the traveled way and replacing traffic lights with separations and interchanges would achieve a directional capacity of 3600 without adding lanes, slightly reduced by the merging action at interchanges. This is an example of the advantages of free-flowing traffic.

CHAPTER 7
INTERSECTION DESIGN

LOW-VOLUME INTERSECTIONS

It is important to assure good sight distance for a low-volume intersection that seldom has more than one car ready to enter at one time. The greatest danger is that two cars will arrive at the same time and the drivers will not see each other. For this reason, all intersections, even simple ones, should be at 90 degrees and never less than 60 degrees. Turn lanes aren't necessary for the lowest-volume intersections, but *slow* signs and intersection signs could warn drivers of possible danger.

When one intersecting road is a minor road, a *yield* sign would alert drivers to prepare to stop because drivers on the major road have right-of-way. Sight distances should be sufficient to give drivers three seconds to see and react, and they should include the stopping distance at the road's permitted speeds. Distance traveled in three seconds' response time include the following:

25 m at 30 km/h	35 m at 40 km/h	40 m at 50 km/h
50 m at 60 km/h	60 m at 70 km/h	65 m at 80 km/h
75 m at 90 km/h	85 m at 100 km/h	85 m at 110 km/h

As traffic volume increases, evaluating volume differences on the intersection's legs will reveal the need for stop signs, as well as where to locate them, either on the lowest volume road or at a four-way stop. As queues begin to form on any leg at the heaviest travel time, the first auxiliary lanes should be right turns because they are the least expensive. This leaves the through and left-turn movements on the original lane. Traffic making a right-turn movement should not have to wait in queue, because the movement is in conflict with only two others: movement on the crossroad straight ahead from the left and left-turning vehicles from the opposite leg. The right-turn lanes should be long enough only to minimize the wait for right-turning traffic. With a low-volume intersection, a minor improvement might suffice for many years. Safety, not capacity, is the principal concern. A sketch of this simple intersection, with minimal right-turn facilitation, is shown in Figure 7.1.

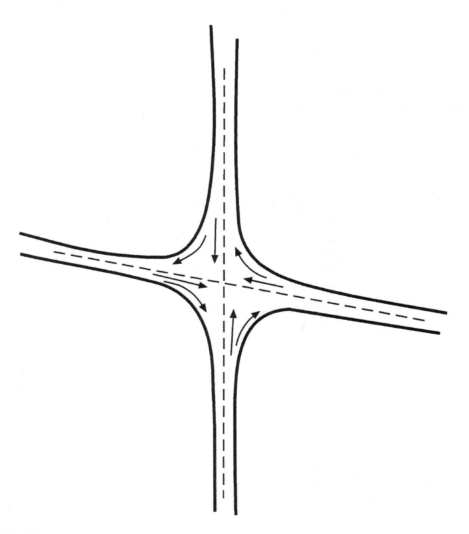

FIGURE 7.1 Four-leg intersection, unchannelized, minimal right-turn lanes.

With the installation of stop signs and continued increase in traffic volume, an intersection will be paved, almost certainly with asphalt. While early paving is likely to be in thin surface treatments, traffic volume and the road's increased importance can justify the use of hot-mix paving at some point.

All road engineers and most drivers are familiar with the effects of stopping heavy traffic on flexible pavement. One part of the solution has been to select a stiffer binder. If Performance Graded binders are available, PG 58 might be the choice for highway pavements; one grade warmer, PG 64, would be the better choice for areas of slow traffic; and PG 70, two grades warmer, is better for areas of stopping traffic. Another concern is to assure dense asphaltic concrete by avoiding segregation of the gravel in the mix. See Chapter 10, Flexible Pavement Design, for more detailed information.

VOLUME-MODIFIED INTERSECTIONS

As volumes increase in the intersection, the movements that cause capacity problems will be evaluated for correction. When the cost of the needed corrections begins to escalate, it is time to forecast future traffic and plan for intersection improvements. Now our concerns are not only for safety and capacity, but also for efficient use of road construction funds. Proper construction will conserve maintenance funds, which will have begun to escalate at this point.

Problems can develop with left turns, right-turn queuing, signage, signals, merging, or lane capacity in each of the three or four legs of the intersection. A suitable geometric plan based on traffic studies and forecasts will meet immediate needs and provide the basic elements, such as right of way or earthwork, for economical solutions. Skimping on these basics can be costly or even preclude future improvements. Because of the analysis of the present and future of each movement in the intersection, a basic intersection type can change from location to location. Proper intersection design and construction can provide an excellent service to the community and save public funds by postponing the necessity of interchange construction for years.

A complex intersection that is designed to handle moderate traffic on the crossroad and heavy traffic on the main road would have a four-phase signal light favoring the main road. Each of the two legs of the main road would enter the intersection with a left-turn lane, two through lanes, and a right-turn lane. Each leg would leave the intersection with two through lanes and a right-merging lane. The two through lanes would merge after leaving the intersection area (Figure 7.2).

The auxiliary lanes for through traffic, right turns, and left turns on the main road will approximate the capacity of the single main-road lane approaching and leaving the intersection area. With 3.6 m pavement and wide shoulders, the main-road lane capacity should be near the maximum 2000 VPH.

The green light on one phase of the four probably would not exceed 40 percent, even when it is favoring the main road. The lane capacity of each of the left and through lanes would be 40 percent of 1800, which is the maximum lane capacity when traffic comes to a stop. With a second through lane, as well as a left-turn lane and a right-turn taper and lane, the signaled intersection will have the approximate capacity of the road that is approaching and leaving the intersection area. See Lane Capacity in Chapter 6, Traffic Estimation and Design Hour Volume.

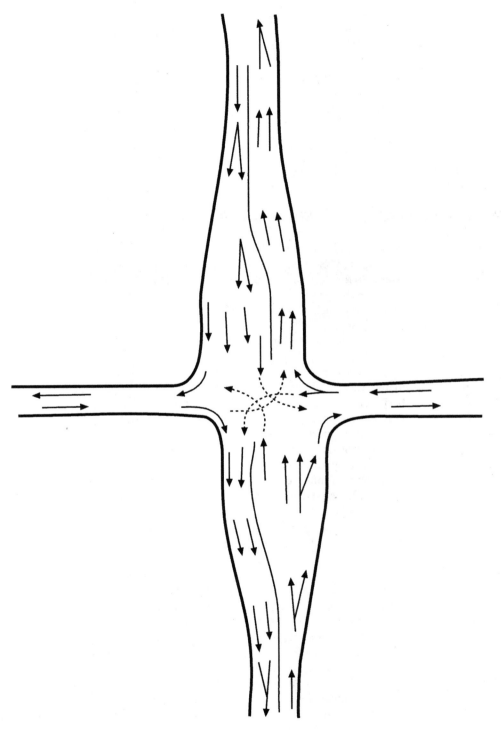

FIGURE 7.2 Complex channelized intersection.

INDIRECT LEFT TURNS AND U-TURNS

When a multi-lane arterial has a median that is too small for a left-turn lane, there are many solutions, each with disadvantages. The designer must choose the solution that best serves the traffic and has disadvantages that are tolerable under the local conditions.

One solution is to prohibit left turns. This forces drivers to turn right and find a loop or U-turn that allows completion of the left-turning movement. This puts a burden on adjacent streets that might be worse than the original problem. This undirected movement is particularly difficult for drivers who are unfamiliar with the street system. Allowing left turns from the left through-lane under heavy traffic causes frustration for drivers and lane changing by the through traffic.

Using the left lane exclusively for left turns, including U-turns, works well for the turns but makes one less through-lane. A small island blocks the lane opposite the left-turning vehicles, leaving almost the whole lane available for turning-movement storage. The remaining through lanes will be free-flowing if there are enough lanes for the through traffic. This might be a good solution for heavy left-turn traffic and many closely spaced intersections.

Ramps on the right work well without reducing the lanes available for through traffic, but they take additional right of way. The jug-handle-type ramp exits before the intersection, and it takes both left- and right-turning traffic to an intersection with the minor road. It will work best in cases where volumes of the turning traffic and of the minor road are not critical. The loop beyond the intersection merges this traffic rather than turning across traffic and it will work better with higher volumes of minor road traffic.

Two arterials can exchange heavy left-turn movements with little disruption of through traffic if one has a wide median for indirect left turns in a crossover. U-turns, although convoluted, are also available. One movement, turning right into the second lane and immediately left into the crossover, will take some driver education. With proper signage and driver experience, this plan can be a solution to this special and difficult problem.

ROUNDABOUTS AND CENTER TURNING OVERPASS

Maryland and other states have begun installing roundabouts, which have long been used in Britain, as a safer design for low-volume intersections. An intersection is replaced with a single-lane circle, which requires somewhat more land but is much safer. (Nearby Washington, D.C. is a fine illustration of the horrors of multi-lane traffic circles!) Each vehicle makes a right turn onto the circle and merges into a single lane of traffic. The circle has merging traffic from the right once for a through movement or twice for a left-turn movement on a four-legged intersection, with perhaps one more merging movement for each added leg of the intersection. Each vehicle exits unopposed and without a merging requirement (Figure 7.3).

Usually, the complete roundabout maneuver can be performed without a stop. Since the roundabout is usually free-flowing, it can save the motorist's time and nerves. Although this design cannot provide the capacity that is available in a complex signalized intersection, it is far less expensive, except in right of way. Its maximum capacity should be near that of a moderately channelized intersection that has far greater construction and maintenance costs. The roundabout is, in effect, a safe and economical (no bridge) interchange for low-volume traffic.

FIGURE 7.3 Roundabout intersection.

The Center Turning Overpass, CTO, is a newly patented development with plans for sale by designer Robert F. Clayton. Designed for street intersections of 35,000 cars or more per day, it is claimed to offer a 40- to 50-percent flow improvement over a standard intersection and does not take up much more space. The right-turn and through movements remain at surface level, while left-turning vehicles enter a ramp and structure to their own intersection above. Two separate two-phase signals increase the green time at both signals. With reported cost comparable to that of a diamond interchange, the CTO might serve some intersections of two major streets better; however, a diamond is best with a major and minor road intersection. The CTO might find a niche where it serves best.

Operationally, the CTO moves traffic as smoothly as does an intersection that prohibits left turns. It still serves the left-turn movement instead of moving it to another intersection or forcing drivers to make a right-turn circle. The left-turning vehicles decelerate on an upgrade and accelerate on a downgrade, which can be steep and short. Clayton has designed standard components, and he plans mass production.

CHAPTER 8
FREEWAY AND INTERCHANGE DESIGN

FREEWAY AND INTERCHANGE WARRANTS

A freeway is warranted when traffic congestion impairs the functioning of the existing road and street system and when delays, vehicle costs, and accidents reach an intolerable level. Other modes of travel, such as plane, train, bus, bicycle, and walking, should be evaluated as partial solutions along with freeway construction. Freeway construction should meet public needs that cannot adequately be served by the other modes of transportation.

Deciding to build a freeway includes making decisions about all road separations and interchanges that are necessary for the freeway's construction and operation. A low-traffic bridge that would not be built as a stand-alone project in the local system can be justified by the public's need to cross the freeway.

The freeway's benefits are the costs saved by road users, including the cost of lost time because of traffic congestion, fuel and other vehicle costs, and the cost of accidents and air pollution. The annual benefits divided by the annual cost of building and maintaining the freeway, including its interchanges and separations, is called the benefit/cost ratio. A ratio greater than one is required to justify the freeway.

The relative benefit/cost ratios of various designs, along with aesthetics, topography, and the degree of service, determine the freeway design. Items considered in freeway design can include the number of lanes, median width and design, frontage roads, road closures and separations, interchange spacing and location, interchange type selection, and interchange ramp configuration.

Many choices must be made in interchange-type selection and design. Each traffic movement must be studied in relation to the available design choices, with attention paid to topography, accommodation of bicycles and pedestrians, and bridge location, size, and orientation. Several designs should be made and compared for benefit and cost.

FREEWAY DESIGN

Freeways are designed for speeds that can be attained in off-peak hours. At peak hours, the traffic volume might approach capacity, and the speed would decline as volume increased. On many urban freeways, off-peak traffic can travel at design speeds, but peak-hour volumes will reduce speeds to near 55 km/h. The design speed should be 110 km/h unless conditions such as mountainous terrain, difficult urban conditions, or expensive right-of-way considerations make a lower design speed advisable. Then the speed should be no lower than 80 km/h. The maximum recommended grades for high-design speeds are shown in Table 8.1.

TABLE 8.1 Maximum Recommended Grades for High Speed Designs

Terrain	Design Speeds				
	80	90	100	110	120
Level	4	4	3	3	3
Rolling	5	5	4	4	4
Mountainous	6	6	6	5	-

Grades can be 1 percent steeper in urban areas and in separate roadways in downgrade. They can be 2 percent steeper for short distances in difficult terrain. Sustained grades steeper than 3.5 percent that include horizontal curves create a dangerous automobile-truck conflict that might require reducing superelevation and lowering speed limits.

Although long, straight freeway sections are desirable, they can become too long in rural areas, contributing to driver fatigue and *highway hypnosis*. Occasional gentle slopes and curves that follow land contours can improve design while reducing earthwork costs.

Medians should be as flat and as wide as possible. In the future, medians can be used for additional highway lanes, mass transit, or fast trains. Future generations might bless us for reserving the space or blame us for not doing it. Providing a wide median can be difficult, which is even more reason for doing it where we can.

The minimum median is 3 m wide. In a four-lane urban freeway, this allows for 1.2 m of shying distance on each side of a 0.6-m median barrier. For six-lane freeways, parking is needed on the left. A shoulder for cars requires 3 m, and a shoulder for trucks requires 3.6 m. With a 0.6-m median barrier, the total median is 6.6 m for cars and 7.8 m for trucks. The truck shoulder is needed when trucks reach 250 Directional Design Hour Volume (DDHV).

Depressed freeways should have these dimensions or greater, as well as a 3.6-m outer shoulder and an additional 0.6-m distance to the retaining wall. The distance can be 0.9 m if the retaining wall has a safety-shaping at traffic level. Vertical clearance should be at least 4.9 m, preferably 5 m to allow for future paving additions. Depressed medians are used frequently in urban areas when traffic is congested, when the right-of-way cost is high, or when there is a need to minimize noise pollution. Depressed medians can be good urban neighbors. They provide access without post-construction intrusiveness, easy reconnection of streets across the freeway, and occasionally room for a park or office building in the air space above the freeway.

Elevated urban freeways are another design option. The elevation can be on fill in high water table areas or in suburban areas with space and less expensive right of way. In urban areas, where existing land uses must be retained, the elevation is on piers. Clearance above surface activities usually varies between 4.5 m and 6 m. Maintenance costs are high, as are costs of construction above existing streets and commerce. Elevated freeways are major bridge structures, and they require individual study and design to fit the conditions of their environment.

GRADE SEPARATIONS

Traffic separation structures are needed where high-volume streets and highways meet, usually with interchange of traffic between them. For high-speed freeways, bridges must accommodate the full width of the roadway, including shoulders across or under the bridge. Although clearance of the full shoulder or parking lane is desirable for other lower-speed roads and streets, area restrictions might justify less clearance, down to a minimum of 0.6 m from the lane edge. Vertical clearance from the finished undercrossing roadway to the bridge should be a minimum of 4.4 m. On interstate highways and most other freeways, the recommended minimum clearance is 4.9 m. To accommodate future resurfacing and to clear occasional overheight vehicles, the desirable clearance is 5 m.

When automobile and railway traffic cross, separation is the goal and benefits both modes. Cost is the only obstacle to complete separation. For safety reasons, we should seek ways to close at-grade crossings, replacing one or several with separation structures. For more information, see Chapter 14, Railroad Crossings.

Grade separations make a major contribution to highway safety. In combination with interchange ramps, grade separations reduce accidents caused by conflicting turning traffic. Separations between freeway interchanges restore access to urban areas. They also reduce the through traffic at interchanges, enhancing the ability of diamond interchanges to handle turning traffic. Closing some road crossings and gathering the cross traffic into a few separated traffic routes will reduce the number of accidents substantially. The closure and separation plans must be made in cooperation with local government and in consultation with local residents. The plans must consider the needs of automotive traffic, bicyclists, and pedestrians.

When a highway crosses a local road, the decision of which crosses over the other can be a complex one. Usually the decision is to favor the road with the greatest traffic, unless topographical and economic considerations clearly favor the other choice. This most often means an overpass. Many reasons support such a decision:

1. The bridge clearance might become obsolete. This has happened on some early interstate highways.

2. Vertical restrictions on the local road would be a lesser problem. Lowering the local road profile, if necessary, is less expensive and less disruptive of traffic.

3. The bridge piers would not be a roadside obstruction on the road with the major traffic.

4. Drivers on the road with major traffic would have unobstructed vision, traffic would have unlimited vertical clearance, and the roadway could reach full capacity.

But some factors favor an undercrossing:

1. If the separation were in an interchange, ramps would be shorter and cheaper because exiting traffic would accelerate on a downgrade and entering traffic would decelerate on an upgrade.
2. An undercrossing reduces cost, as does the smaller bridge and slightly less right of way.
3. A road on an easier gradient near ground level will better serve major traffic in an urban area.
4. A depressed roadway would be better for noise control, would make crossing easier, and would reduce cost. Some of the airspace above the roadway might be used.

DIAMOND INTERCHANGES

The diamond interchange is nearly always the first choice when separating traffic of a major road and a minor road. It provides for uninterrupted through traffic on the major road. Traffic flow on the minor road is freed from cross traffic conflicts but not from conflicting turning traffic. Traffic turning from and to the major road is provided with ramps, acceleration lanes, and deceleration lanes. Traffic turning from and to the minor road is served at intersections on each side of the major road. A simple diamond interchange design is shown in Figure 8.1.

Design for the ramps' intersection with the minor road follows the principles described in Chapter 7, Intersection Design. The ramps are one-way roads, and the through movement from the off-ramp to the opposite on-ramp might or might not be provided. Through traffic would bypass accidents on the overpass bridge but would reduce intersection capacity for normal traffic.

This inexpensive type of interchange can handle a high volume of traffic: 2200 vehicles per hour per lane on the major road, with 1800 v/h/l on the minor road, minus the capacity lost because of turning traffic at the ramp intersections. As noted in Chapter 7, this directional volume capacity on the minor road can be continued by adding lanes within the interchange area, even after ramp volumes require signalized intersections.

To keep bridge reconstruction from disrupting highway use, the separation structure should provide the space that will be needed by the minor road during the lifetime of the bridge. This usually is 50 years or more after the date of project completion. Destroying a good bridge to make room for more lanes under it is an example of poor planning. Proactive planning and construction are far less expensive than having to solve the problem of construction under heavy traffic. The other interchange elements should be designed for 20 years or more unless special circumstances apply.

If necessary, phased construction can control early costs. For example, fifth and sixth outside highway lanes can be constructed through the earthwork phase but left unpaved until later. Usually, space for the lanes will be left in the median with or without earthwork to avoid greater cost and traffic disruption at the interchange ramps. This phasing should not affect bridge planning and construction, but a bridge might be planned to

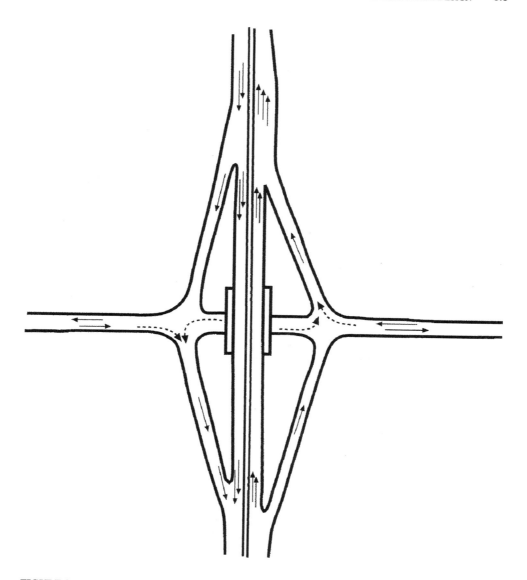

FIGURE 8.1 Diamond interchange design.

serve one roadway for both traffic directions initially, with space for another bridge reserved until a parallel roadway is needed. Planning of such phased construction should consider interest savings for several years and possible acceleration of other priority projects, as well as the added cost of working near traffic and the possible inflation of material costs.

MODIFIED DIAMONDS

A diamond interchange is modified when one or more, but not all, movements differ from the standard diamond configuration. An example is shown in Figure 8.2, where the northbound to westbound left-turning movement is assumed to be near the maximum capacity of the diamond configuration soon after opening to traffic. To increase capacity for the movement, a loop to the right, more typical of the cloverleaf interchange, is provided. It replaces the left turn into the northwest quadrant and moves this traffic from the right lane of the minor road to a westbound acceleration lane. Because the left turns are the greatest impediment to free flow of traffic in the diamond design, a loop for the left turn that develops heavy traffic first is a likely modification.

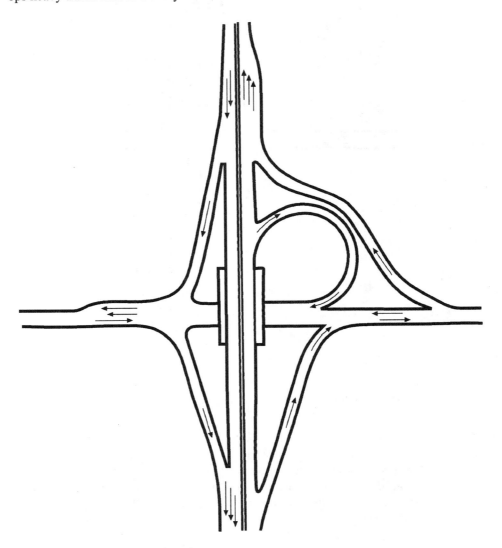

FIGURE 8.2 Modified diamond interchange.

With the loop, the bridge will be longer to accommodate one extra lane. Also, the directional ramp in that quadrant must be designed to avoid the loop, and the design requires much more right of way in the quadrant. It increases the capacity of the minor road, however, by eliminating the greatest restriction of that capacity, the high-volume left turn. This makes more green-signal time available for other movements.

If the other left turn off the minor road is similarly designed in a loop, the modification has progressed halfway to the cloverleaf interchange. The two ramp and minor road intersections would be farther apart and all right turns would remain (plus two left turns converted to right turns). There would be no left turns *off* the minor road but still two left turns *onto* it. This type of diamond interchange modification might be chosen when traffic forecasts of some of the left turns approach the limits of the simple diamond design, even with complex traffic signals at the intersections.

A similar modification would eliminate the two left turns for traffic exiting the main road. If all four left turns are provided in loops, the interchange has become a cloverleaf. If a diamond interchange has been modified with one to three loops, it might be called either a modified diamond or a partial cloverleaf.

The main road's exits to the minor road can be modified for heavy traffic in many ways. The ramp can be widened to two, three, or even four lanes just beyond the exit, providing more storage for the right and left turns at the intersection. The exit can be moved farther from the structure to permit more storage in the exit lanes. A lane, or even two, of any length can be added to the main road for exit only. Backup of stopped exiting traffic on the major road through lanes must not be permitted. This would be an unacceptable traffic hazard.

As noted in Chapter 7, intersections with heavy turning movements have much less traffic capacity than the road away from the intersection unless lanes are added in the intersection area. Two closely spaced intersections with a bridge between them at a diamond interchange indicate possible early obsolescence of the bridge. It would be wise to estimate 50-year traffic on the crossroad and to evaluate of the possibility of widening it in 20 to 50 years.

Another diamond interchange modification could be elimination of one or more of the eight traffic movements. Paired movements usually would remain, and the eliminated movements should be available nearby. An example would be two important but minor parallel roads that are less than a kilometer apart and have a good road network between them. The design solution might be two half-diamond interchanges that avoid the unsafe practice of placing two ramps too close together (Figure 8.3).

Another example would be a location where an interchange is needed, but the right of way in one quadrant is so expensive (a factory is located there, for example) that it vetoes the project. The decision might be to build a half or three-quarter interchange.

The single-point diamond interchange (Figure 8.4) is the most modern modification, dating from the early 1970s. Structured over or under the main road, as required by the site, all left turns and through movements on the minor road are gathered into a single intersection that requires only a two-phase signal. Right-turn movements are free-flowing and directional. Both left- and right-turn ramps can be multi-lane, as required by projected traffic volumes. The single-point diamond requires a little more right of way than a simple diamond does, and it handles much greater traffic volumes, but construction cost is high. It is, therefore, the choice for high-traffic urban locations and other sites that have heavy traffic and right-of-way restrictions.

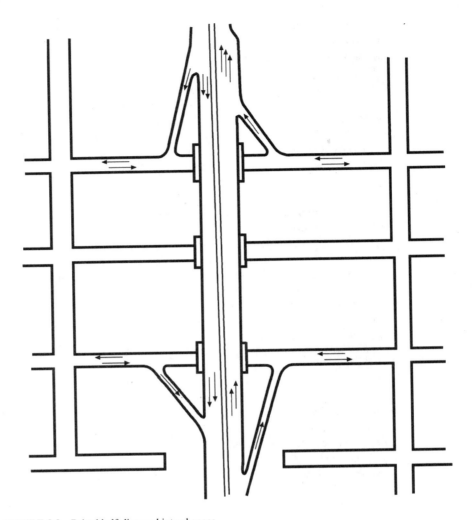

FIGURE 8.3 Paired half-diamond interchanges.

CLOVERLEAF INTERCHANGES

A cloverleaf interchange, as shown in Figure 8.5, is the simplest interchange that has no restrictions on through traffic and no stops or intersections for turning traffic. It allows for the construction of any needed number of through lanes on either of the roads. It is an economical interchange, usually needing a single bridge for traffic separation. It converts each of the four left-turning movements into a right-turning loop, preferably with a 45-m radius. In tight quarters, however, the radius can be as small as 20 m, or compound curves with spirals might be used. The time lost because of reduced speed on the tighter circle is in part regained because there is a shorter distance to travel on the reduced circumference.

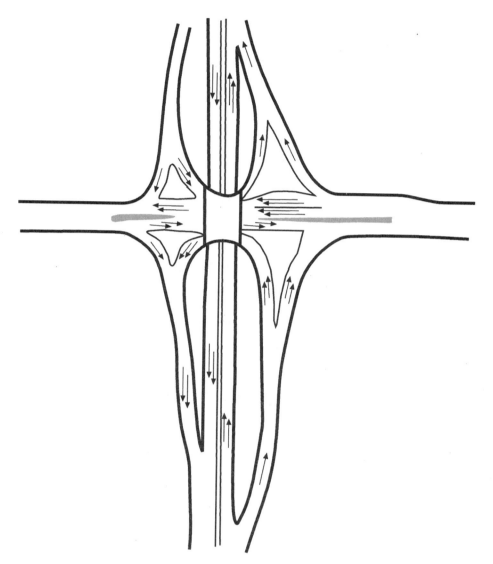

FIGURE 8.4 Single-point diamond interchange.

A partial cloverleaf interchange can be one with some diamond-type ramps and some left turns in clover circles. A three-legged interchange with the three directional right turns and the one left turn in a loop (Figure. 8.6) is a special form of cloverleaf, usually called a trumpet. It works well and is economical, but it should not be built unless the fourth leg is unlikely to be needed. Trumpets are very difficult and expensive to modify. Bridge relocation might be required, and traffic disruption would be extreme.

FIGURE 8.5 Cloverleaf interchange.

Three-legged interchange designs, with the fourth quadrant to be added later, can be built at reasonable cost and cause moderate traffic disruption (Figure 8.7).

The cloverleaf has several disadvantages. On each side of the bridge, traffic accelerating from a loop must merge with decelerating traffic headed for the next loop. This weaving movement is dangerous, especially under heavy traffic. Except in cases of low traffic, the designer usually chooses to add a separate lane, called a collector-distributor, for this ramp merger. The collector-distributor road or lane reduces the hazard but hikes cost by widening the bridge. It is important to begin the exit lane well before the merger with the entering ramp and to continue it beyond the exit ramp before merging with the through lane.

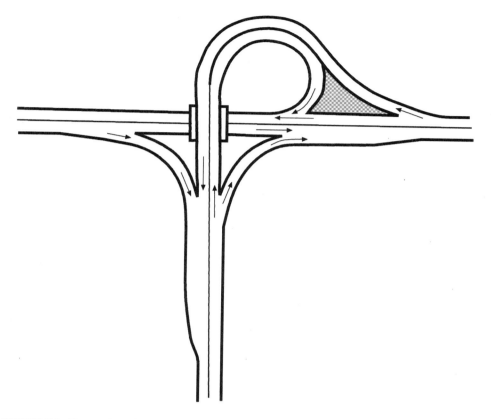

FIGURE 8.6 Trumpet interchange.

The out-of-direction loop reduces operating speed if it is a tight radius, and it requires much more right of way if it is not. The loops and the right-turning movements outside them assure that the design is not economical of right of way. A right-turn movement might be two lanes if traffic volumes or forecasts require it; this is not so for the left-turning movements served by the loops.

More than one lane is difficult to design on circular ramps and probably will not achieve much additional capacity. A second lane on a loop adds little except storage capacity, and it might cause traffic-weaving problems.

As an early design for the meeting of two major highways, the cloverleaf served well for years. Although it is displaced by directional interchanges when ramp volume projections exceed loop capacity and crowded on the low-volume side by the diamond interchange design with improved intersections and modifications, the cloverleaf still fills a niche. It fits a junction of two major highways better than the diamond does. It also costs less than a directional interchange if right of way is inexpensive and if projected traffic on the circular ramps in the design year will not exceed the one-lane limit of 2200 v/h.

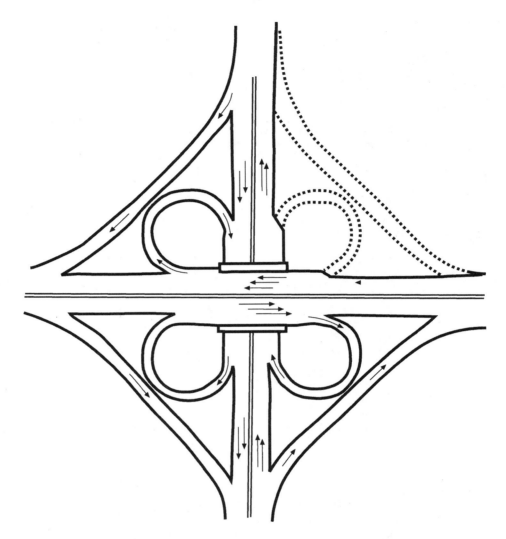

FIGURE 8.7 Three-legged interchange.

DIRECTIONAL INTERCHANGES

When increasing traffic loads at freeway junctions outran the cloverleaf ramps' capacity, the solution was to take each ramp directly to where the traffic wanted to go, building all necessary bridges. Thus was born the *directional* interchange, which has bridges over bridges, sometimes four stories high (Figure 8.8).

Soon modified, a modern directional interchange is a collection of compromises, like most highway designs. Where feasible, ramps align to reduce bridge construction, and

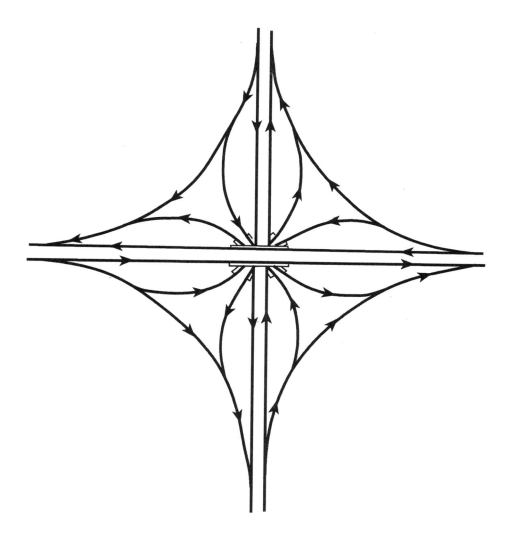

FIGURE 8.8 Directional interchange

bridges align to moderate their length as they cross other roadways. Considered by many as aesthetically displeasing—some call them "a plate of spaghetti"—directional interchange designs have become more acceptable, but they will never rival Greek statuary.

More important, they did, and do, their job. Ramps can be any number of lanes. The design can include three to six legs of intersecting highways. Through lanes are unimpeded in the interchange area. Merging distances are usually ample and always adequate. Although it might look like a tangle in plan or from the air, to the driver who sees little more of it than the route ahead, it looks direct and takes that driver to the desired destination with easily understood signage.

For a T or Y intersection, a directional interchange is designed easily, saves space and funds, and is excellent operationally. It usually is the design choice, unless it is probable that the interchange's fourth leg might be built.

A modified directional interchange can have one or more circular ramps where traffic volumes are too low to justify a directional ramp. A semi-directional interchange has one or more movements less than direct but is not as indirect as a loop.

As in cloverleaf interchanges, directional designs should avoid weaving sections by redirection or by collector-distributor roads. Left exits, when unavoidable, should be clearly marked and have warning signs far in advance of the movement to allow room for lane changes across the roadway. Left entrances, when unavoidable, should be on added lanes that are continued a reasonable distance for merging before lane closing.

Directional interchanges are becoming the clear choice for most freeway junctions as traffic volumes escalate and future volume projections go higher and higher. Although directional interchanges are expensive to construct when compared to diamond and cloverleaf interchanges, they are a part of the cost to serve the desires of automobile drivers in a high-traffic era.

CHAPTER 9
PAVEMENT BASE

SUITABLE SUBBASE MATERIAL

In 1970, the American Association of State Highway and Transportation Officials (AASHTO) defined clay as having particles below 0.0002 mm, passing the Number 200 screen. Silt is defined as having particles between 0.0002 mm and 0.075 mm, retained on the Number 200 screen. Fine sand has particles between 0.075 mm and 0.425 mm, retained on the Number 40 screen. Particles of coarse sand are between 0.425 mm and 2 mm, retained on the Number 10 screen. Particles of gravel are between 2 mm and 75 mm, retained on the Number 2 screen. Any material with particles above 75 mm is defined as rock.

Usable subbase soils will bear heavy loads by being well-graded in texture; in other words, the soil's grades will fit together under compaction. The subbase may be open-graded with a compact, water-resistant subgrade draining water to the ditch, or it may be dense and water repellent to protect the subgrade. A usable open-graded subbase soil must be coarse enough (low enough in fines) to drain excess water. An excellent coarsely graded mix for subbase might test 50 percent retained on Number 10, coarse sand, gravel, and loose rock; 20 percent on Number 40, fine sand; 15 percent on Number 200, silt; and 5 percent clay passing Number 200. This drainable mix has little fine and very fine material, leaving voids to facilitate drainage and eliminating excess subgrade moisture that could cause a loss of strength. Such excess moisture is frequently found in hillside cuts, where subsurface flow can be intercepted. Cross culverts to the main water sources, perhaps fed by a deepened ditch or underground drain of 2-in. slotted PVC pipe, would be one way to control this. Water control under the pavement is essential because the freeze-thaw cycle can destroy the pavement quickly. Water control is far less expensive than other methods of prevention, such as removing expansible material to the frost line or using coarse material layers, insulation, geothermal materials, or thicker pavement.

A well-graded mix for subgrade has coarse, medium, fine, and very fine material that is resistant to water penetration. Under compaction with optimum water content, it becomes very dense and able to bear heavy loading. This material is very good for the top 0.3 m of the subgrade, and it is able to provide strong support for the subbase, base, and surfacing. Both well-graded and coarse mixes are useful as Selected Roadway Borrow (SRB) to replace or amend unsuitable materials in the fill. Very fine material, such as clay, loam, silt, and very fine sand, should be no more than 35 percent by weight of samples and no more than 25 percent of the top 1 m of the fill. Pockets of very fine material should never occur in fill or cut under the pavement, as they can liquefy and drain away, leaving a void under the pavement that will become a pothole. When such very-fine-material pockets are found, they must be blended with other material in the top 0.3 m of subbase by plowing, raking, and recompacting.

Organic material is unsuitable for subbase and fill, except on fill slopes where it might act as mulch to aid slope seeding. When fill is unsuitable for the subgrade to support the pavement, SRB can be hauled from any location within reasonable distance or from a borrow pit. Stockpiling suitable SRB from the cuts instead of using it deep within fills can save money. SRB can be used instead of fill material on the top 0.3 m of the roadway, or it can be used as an amendment if it changes the composition of the fill material enough to make it suitable for subbase. As an amendment, SRB would be spread on top of the fill, plowed and raked to mix the top 0.3 m, and compacted. Cuts also can have material that is not suitable for subgrade support. Such material must be replaced or amended.

An excellent finely graded mix for fills could be 50 percent coarse sand retained on Number 40; 25 percent fine sand retained on Number 200; and 25 percent silt and clay passing Number 200. A good fine-sand material could be graded 51 percent or more coarse sand retained on Number 40; 10 percent or less fine sand retained on Number 200; and 0 to 39 percent silt or clay. (See Table 9.1, AASHTO Soil Classification.) A good mix of silty gravel and sand for fills could be 65 percent gravel and sand retained on Number 200 and 35 percent fine sand and silt passing Number 200. A fair mix of clayey gravel and sand would test the same, except the plasticity index would increase from 10 to 11, indicating the presence of clay rather than silt. A poor mix would differ from a fair one in that the percentage passing the 200 screen would be 36 percent or more. Peat and other highly organic soils are classed as A-8. This means they are unsatisfactory for construction purposes but useful for landscaping.

Although material passing the Number 200 screen is defined as clay, it actually is a mixture of clay, fine silt, very fine sand, and organic material. The actual content can make a difference. Fine sand is a good construction material, fine silt is fair, clay is poor, and organic material is bad. The dispersion test is the quickest way to identify the proportions of fine material passing the Number 200 screen, which would have particles less than 0.075 mm diameter. Thoroughly disperse (dissolve) a measured sample. After letting it settle for one minute, pour off the remaining liquid. The sediment is sand. After about 60 minutes of additional settlement, pour off the remaining liquid. The sediment is silt. Clay will remain in solution a long time, and it can be assumed to be the remainder of the sample. If clay measurement is desired, the water must be boiled off the sample. Dark-colored organic matter, usually loam, will be mixed in the clay and silt. Silt is usually tan to gray, and clay is usually tan to orange. Loam is usually medium-brown to

TABLE 9.1 AASHTO Soil Classification

Soil Mix	Rating	Description	
A-1-a	Excellent	Retained on #10, gravel & rock	50% or more
		Retained on #40, coarse sand	20% or less
		Retained on #200, fine sand	15% or less
		Passing #200, silt and clay	15% or less
		Plasticity Index	6 or less
A-1-b	Excellent	Retained on #40, rock, gravel, sand	50% or more
		Retained on #200, fine sand	25% or less
		Passing #200, silt and clay	25% or less
		Plasticity Index	6 or less
A-3	Good	Retained on #40, coarse sand	49% or less
		Retained on #200, fine sand	41% or more
		Passing #200, silt and clay	10% or less
		Non-plastic fine sand	
A-2-4	Good Silty Gravel/Sand	Passing #200, silt and clay	35% or less
		Liquid limit 40 Max. Plasticity Index 10 Max.	
A-2-5	Good Clay Gravel/Sand	Passing #200, silt and clay	35% or less
		Liquid limit 41 Min. Plasticity Index 10 Max.	
A-2-6	Fair Silty Gravel/Sand	Passing #200, silt and clay	35% or less
		Liquid limit 40 Max. Plasticity Index 11 Min.	
A-2-7	Fair Clay Gravel/Sand	Passing #200, silt and clay	35% or less
		Liquid limit 41 Min. Plasticity Index 11 Min.	
A-4, A-5	Fair Silty Soils	Passing #200, silt and clay	36% or more
		Plasticity Index	10 or less
A-6, A-7	Poor Clay Soils	Passing #200, silt and clay	36% or more
		Plasticity Index	11 or more

black. Measure sand and silt after heating or after overnight draining and air-drying. Your answers will not be exact. This flour clings to rock, gravel, and sand, and the settlement test is imprecise. However, you will have better information on the actual amounts of fine sand and silt present, and you might change your opinion about the sampled soil.

Beyond the sieve analysis, soils are compared by liquid limit wl and plasticity index Ip. In the field, nuclear gauges measure the moisture content and density of undisturbed cuts and compacted fills. In the modified Proctor laboratory test, controlled compaction determines the dry density and optimum moisture content in clay soils.

At 100-percent compaction, when all air voids are eliminated, the maximum dry density PD and the optimum water content w are obtained.

Compaction at 100 percent Proctor is seldom attained in the field, and plans usually specify a percentage of Proctor density. Table 9.2, Moisture and Compaction, provides recommendations for these percentages to a 1m depth of fill and for deeper fills, ranges of optimum water content to achieve maximum density, and the attainable range of dry density for various classes of soil.

TABLE 9.2 Moisture and Compaction

Class and Type of Material		kg/m3 Dry Density	% Water	Suggested % Proctor Max	
				Top 1m	Below 1m
GW,	Well graded clean gravel and sand	2000-2160	11-8	97	94
GP,	Poorly graded clean gravel and sand	1840-2000	14-11	97	94
GM,	Poorly graded gravel/sand/silt	1920-2160	12-8	98	94
GC,	Poorly graded gravel/sand/clay	1840-2080	14-9	98	94
SW,	Well graded clean sand and gravely sand	1760-2080	16-9	97	95
SP,	Poorly graded clean sand and sand/gravel	1600-1920	21-12	98	95
SM,	Silty sand, poorly graded sandy silt	1680-2000	16-11	98	95
SM-SC,	Sand/silt/clay with plastic fines	1760-2080	15-11	99	96
SC,	Clayey sand, poorly graded sandy clay	1680-2000	19-11	99	96
ML,	Inorganic and clayey silts	1520-1920	24-12	100	96
ML-CL,	Organic silt and clay	1600-1920	22-12	100	96
CL,	Inorganic clay, low to medium plastic	1280-1600	24-12	100	96
OL,	Organic silt and silt/clay, low plastic	1280-1600	33-21	-	96
MH,	Inorganic clayey silt, elastic silt	1120-1520	40-24	-	97
OH,	Organic and silty clay	1040-1600	45-21	-	97

Clay soils can be solid, plastic, or liquid depending on the water content. When clay is compacted by the weight of 1 m or more of fill, it should remain solid and loadbearing unless water frequently penetrates the fill. The water contents at these transitions are the plastic limit wp and the liquid limit of the soil. The Ip is the difference between the liquid limit and plastic limit.

$$Ip = wl - wp.$$

The Atterburg Limit Test is a consistency test for liquid limit. Two parts of a moist sample are separated a specific distance and the container is dropped 25 times. This is repeated with added water if the sample is still separated. At the liquid limit, the parts of the sample will be touching for a length of 13 mm. In the test for plastic limit, a soil sample is rolled into a thread that is 3 mm in diameter. If it crumbles at that diameter, the plastic limit has been reached. These limits are not of significant use for fine sand, and the plastic limit is of little use for silt. The liquid limit is significant for silt and clay, and the plastic limit is significant for clay.

The A-1-a and A-1-b soil mixes are excellent for building roads, including the upper portion that supports the pavement. They can be used as an amendment to improve the poor silty and clayey soils graded A-4, A-5, A-6, and A-7, which might be unsatisfactory for fill without amendment. They also can be used to replace or amend the upper 0.5 to 1 m of the fair A-2-6 and A-2-7 soils to make them satisfactory for pavement support.

The R-value of a soil is a measure of its ability to resist lateral deformation when a vertical load is applied. It is determined in the laboratory by testing saturated, compacted samples. Since the samples are saturated, the R-value is the worst performance to be expected from the material. R-values run from 0, the strength of water, to 100, the strength of steel. Road materials run from R-5, wet clay, to R-85, hard rock. In the absence of a laboratory stabilometer test, engineers in the field can estimate R-values with screen tests, observation and experience with soils, using the parameters listed in Table 9.3.

TABLE 9.3 R-Value Ranges for Road Building Material

Material	R-value
Good crushed rock	75-85
Gravel	20-85
Clayey and silty gravel	15-70
Sand	30-80
Silty sand	10-80
Silt and sandy silt	10-60
Clayey sand and clayey silt	6-42
Sandy clay	7-22
Clay	5-15

At first, the spreads appear to be too wide to do much good. Looking closer, it is apparent that the highest numbers are for sharp material: crushed rock, broken gravel, and sharp sand. The lowest numbers are for material that is heavily loaded with clay, with silt-heavy material faring only slightly better. The middle numbers are for fairly clean and smooth material. Now we can subdivide the classes such as clayey and silty gravel shown in Table 9.4.

Most engineers should make close estimates within these subgroups using screening tests and the feel of the material. Experienced engineers might be able to estimate an R-value close to the actual value found by a stabilometer test.

When poorly performing subbase and base must be used, the pavement depth must compensate. The necessary depth can be derived from charts in AASHTO's Guide for Pavement Design; however, if the strengths of subbase and base total 40,000 psi or more, engineers should base the pavement depth on other criteria. These criteria might be the available aggregate, desired durability, and surface texture, or other such factors.

TABLE 9.4 R-Values for Clayey and Silty Gravel

Material	Upper 1m of roadway	Deeper than 1m in the fills
Pea gravel, heavy with clay	15	30
Pea gravel, heavy with silt	25	45
Sharp gravel, heavy with clay or silt	35	60
Pea gravel, little clay and silt	40	55
Sharp gravel, little clay and silt	55	70

POZZOLAN AND ASPHALT AMENDMENTS

If SRB is not available to amend the soil in the top 0.3 m of the fill or cut, a more expensive alternative is to purchase pea gravel, crushed stone, or other material that will supply the missing grades in the mix. A less expensive solution might be to use a pozzolan. Pozzolans are cementitious materials that combine with sand and silt to produce a solid capable of resisting downward force, displacement, and water penetration. These become a very dense subbase, protecting the subgrade and carrying the water penetrating the pavement cracks into the permeable base to the ditch. Locally available pozzolans that might be less expensive than portland cement are lime, fly ash, diatomaceous earth, and silica fume. Clay is a pozzolan, but its proper dispersal and moisture content cannot be assured under field conditions. Its presence, however, will reduce the amount of other pozzolan needed.

Depending upon its cementitious action with the native material, pozzolans or asphalt can be used as an amendment instead of portland cement. This will make the native material in the top 0.3 m of the subgrade suitable for bridging to protect the pavement from the deterioration of the subgrade. To assure pavement viability in all cases, subgrade movement must be minimized by removing water from the subbase, or base, into the ditch.

PORTLAND CEMENT CONCRETE BASE

When other choices are not economically feasible or would not protect the pavement from voids or frost in the subbase, the remaining solution is to add strength to the pavement by using thicker or pre-stressed pavement. See Chapter 10 for more information.

Laying a portland cement concrete (PCC) base might be more economical than thickening the pavement. The base might be stronger, and it needs no reinforcement and no finishing. The cost of two passes, rather than one, of the paving machine would balance these savings.

An alternative method that costs less but offers less control is on-site mixing with road machinery. It is best to do this when clay content is less than 35 percent in classes A-2-6 and A-2-7. The surface should be scarified to a depth of 0.15 m to 0.3 m. The portland cement spread should be 6 percent (for 35 percent clay) to 12 percent by volume, and the material should be bladed to mix and smooth, brought to optimal water content, and rolled.

CHAPTER 10
RIGID PAVEMENT DESIGN

PAVEMENT DESIGN AND CONSTRUCTION

Portland cement concrete (PCC) pavements have greater durability and longer service lives than pavements made of other common materials. Other advantages of a PCC pavement are the way its color contrasts with asphalt or gravel shoulders and the way it resists water damage unless it is undermined. Its primary disadvantage is that it has a greater initial cost than flexible pavements.

All paving materials need well-compacted subgrade that is well drained and made of material capable of supporting the pavement. PCC pavement bridges small voids better, but asphalt is easier to repair. All paving can expand and buckle under extremely high summer temperatures. While PCC is least likely to do this, the damage is more serious when it does.

Pavements lose texture with use, especially in locations where tires exert great force because vehicles are starting, stopping, and maneuvering curves. Because of the higher speeds and possible low superelevation on curves, lateral forces can polish the pavement surface so much that highest-speed traffic is in danger of being pushed off the pavement by centrifugal force. PCC is more resistant to polishing than other materials, but correction is more costly and involves diamond sawing of parallel cuts to provide renewed traction. Starting and stopping traffic exerts more force on the pavement, but this is seldom a problem on PCC because of its great compressive strength.

MIX DESIGNS

Portland cement is a dry mixture of finely ground lime, silica and alumina, and a small amount of plaster of Paris. The quantities can vary according to the type of cement desired. Of the five main cement types; two are important in massive concrete construction projects, and the following three are important in road construction:

Type I. Normal portland cement, the most common cement, is used in pavements, beams, culverts, and sidewalks.

Type III. High early-strength portland cement is used when the structure is needed quickly or when cold weather limits curing time. Type III cement has a high shrinkage rate and cracks more often.

Type V. Sulfate-resistant portland cement is used where sulfates, such as those found in very alkaline soils, are present.

PCC is produced when portland cement is mixed with water and an aggregate of rock, gravel, sand, and other materials. The cement and water enter into a chemical reaction and harden, forming a binder around the particles of aggregate. A mixture that has just enough water to produce the chemical reaction is very harsh and hard to work, and it is difficult to assure that all the cement will get wet. Too much water in relation to the cement, however, weakens the concrete. The strongest concrete has just enough water to make a workable concrete mix. Additives that lubricate the mix will reduce the amount of water needed without stiffening the mix. Vibrating the mix after it is placed forces the cement-water mortar into the voids between particles and allows a stiffer mix to be used. Both of these methods increase the strength of the concrete.

Portland cement is PCC's most expensive major component. An economical mix, called a well-graded mix, has a minimum number of voids to be filled with the cement-water mortar. The most economical mix would start with a maximum rock size of usually one-half the slab's thickness and less in reinforced slabs. Large rock might reduce workability and require more water that would weaken the mix. After selecting the largest rock size, smaller sizes of rock, gravel, and sand are selected to fit together to leave minimal voids for the hydrated cement to fill.

In addition, the mix might be made more economical by substituting locally available and less expensive pozzolans, such as lime, fly ash, silica fume, or diatomaceous earth, for part of the portland cement. Other admixtures can meet the following purposes:

Damp-proofing to resist mold and mildew

Corrosion inhibitors to reduce attacks on steel reinforcing

Polymers to bond, reduce permeability, and promote resistance to abrasion and the freeze-thaw cycle

Air entrainers to provide freeze-thaw resistance and increase workability

Accelerators to reduce setting time for cold-weather work or early traffic.

A truncated cone called a slump cone measures the mix's workability. The cone is filled with the mix, excluding the largest aggregate. As it is placed in the cone, each one-third of the concrete is carefully compacted with a round rod for density. The cone is gently lifted and the sample's height is compared with the cone's height. The difference is called the slump, which is the measure of workability.

A low slump, such as 25 mm, indicates a stiff mix and strong concrete. If more water is needed for workability, it should be added a little at a time, and the slump test should be repeated. A slump of 25 mm is the minimum slump that concrete should have. Concrete for pavement should have a preferable slump of 25 mm and a maximum slump of 75 mm. Concrete with 100-mm slump is acceptable for heavily reinforced walls, slabs, beams, columns, and pavement subbase. Concrete with slump higher than 100 mm should be rejected for any construction purpose.

Several cylinders are cast from each mix and used for strength testing. They are set aside for curing, and a load is applied to a cylinder from each mix for the desired curing time. The load is increased to the point of failure, and the compressive strength is calculated by dividing the cross-sectional area of the cylinder into the load's size at failure. Concrete increases in strength for many years; a practical working strength is acquired in 28 days. Compressive strengths most often vary between 3000 and 6000 psi, but strengths as low as 2000 psi or as high as 8000 psi are not unusual. Because tensile strength of concrete is approximately 8 percent of compressive strength, concrete is seldom used to resist tension unless it is prestressed.

Normal PCC averages 94 kg/cm^3 and varies by 6 kg/cm^3. The type of aggregate used is the primary weight factor, but aggregates can vary considerably because of different pour purposes. When high strength is not needed, cinders and vermiculite produce lightweight concrete. Using steel balls for aggregate in concrete counterweights will produce a concrete approaching the weight of steel.

To minimize cracking, reinforcing bars (called *rebars*) or mesh, can be placed ahead of the slip-form paving machines, but reinforcing is not required. If the concrete is not reinforced, construction joints spaced at 30 times the thickness of the concrete should be sawed in the green concrete to localize cracking. This should be done at 4.5 m for 0.15 m pavement, 6 m for 0.20 m pavement, and 7.5 m for 0.25 m pavement.

If the slab is reinforced, joint spacing should be between 15 m and 30 m, depending on the amount of reinforcing rather than on the pavement depth. The steel required is between 0.5 percent and 0.8 percent of the pavement cross-section. Rebars must resist pulling out of the concrete. ACI 31812.2.1 specifies a length, called the development length, that will prevent pulling. Using hooks, a greater amount of cover, or a higher percentage of steel in the pavement cross section, can reduce the development length.

When the machine stops for an extended period, 0.45-m dowels at 0.30-m spacing will provide sufficient development length. Primarily, they will resist shearing force in load transfers between slabs. Bars should be number-6, 20-mm diameter for 0.15-m pavement thickness, number-8, 25-mm diameter for 0.20-m pavement, and number-10, 30-mm diameter for 0.25-m pavement.

Expansion joints also will resist shearing forces, but there must be room on each side of the joint for lateral movement of the pavement caused by temperature changes. Alternating ends of the rebar dowels should be coated with material that will prevent bonding of the concrete and steel.

CONTINUOUS SLAB

After PCC is placed (usually at a slump of 25 mm) ahead of a slip-form paver, the paver and its train spread, vibrate, screed, and finish the pavement without stopping.

This machine does for road construction what the combine did for wheat farmers!

In addition to saving labor and providing road-user comfort, the paving train does a much better job than is possible by manual methods. Less slump means stronger concrete, and a continuous screed and mechanical float and finish make a smoother surface.

PRESTRESSED CONCRETE PAVEMENT

Prestressed concrete pavements first were built in European countries in the 1950s, and they have experienced an expected 30 years or more of service life. They never have been in general use in the United States. American demonstration projects, as well as European experience, indicate that they have competitive initial costs and reduce pavement maintenance.

A pre-stressed pavement is far stronger than a standard pavement. PCC has tensile strength that is only about 8 percent of its compressive strength. Prestressing produces a slab that is 50 percent thinner but has high compressive and tensile strength. Prestressed pavements require steel that is 0.1 percent to 0.2 percent of the pavement cross section: about 20 percent of the steel in continuously reinforced concrete pavements. With joints spaced at 100 m to 200 m and careful joint construction, the ride should approach that of the continuous slab.

CHAPTER 11
FLEXIBLE PAVEMENT DESIGN

SUBGRADE AND SUBBASE CONSTRUCTION

Compared with PCC, flexible bituminous pavement is much more dependent on base strength and uniformity. Small pockets of subgrade weakness that PCC might bridge easily would cause depressions under asphalt cement pavement (ACP), and depressions can develop quickly into potholes. The subgrade is the best grade of native soil reached in cuts and reserved for the upper 1 m of adjacent fills. It also can be used to replace or amend material that is not suitable for supporting the weight of repeated heavy wheel loads in the upper 0.3 m of finished cuts. As detailed in Chapter 9, fine-graded material in particular must not exceed 35 percent of the subgrade's volume, although such material might be suitable for fill below the top 1 m. If the subgrade has sufficient strength and density, it also can serve as subbase. Another alternative would be a subbase made of a compacted, well-graded material placed on the subgrade.

A strong subbase accepts and distributes traffic loads from the pavement base and pavement. A dense subbase keeps water from rising from the subgrade, protecting the base and pavement from frost damage and from clay and silt intrusion. The sloped, densely compacted subbase surface drains the water that penetrates into the granular base through pavement cracks. Careful subgrade investigation, as well as careful subgrade and subbase design, can assure the pavement's strength, density, and drainage capacity, avoid early failure, and achieve the life cycle that flexible pavement can provide.

BASE AND PAVEMENT CONSTRUCTION

The base course, which directly supports the pavement, is the first course with precisely specified content. It consists of crushed stone or slag, gravel, and sand. Limiting fine particle sizes creates voids in the compacted base. These voids, and an impermeable subbase or subgrade, provide a drainage path that carries water that penetrates under the pavement into the ditch. The materials' strength, stability, and hardness can be specified. Admixtures of portland cement, asphalt, or lime can stabilize the materials and make them satisfactory for the base course.

The base can be placed in multiple courses, and each would have separate specifications. The first course would have larger aggregate sizes and the greatest voids so it could provide water channels to the ditch. The upper course or courses would have the role of pavement support and would contain fewer voids, more fine material, and more stabilizing admixtures. A base course's minimum depth should be 100 mm or twice the largest aggregate grade, whichever is greater. The number of courses used should be the fewest needed to achieve the design objectives, because time and labor costs are directly related to the number of courses that are laid.

ACP aggregates and their grading are similar to those used in rigid pavement, except that broken aggregate faces are much more important to ACP. Specifications for the ACP base require a percentage of broken rock planes, because material keyed together under compaction adds to the strength and stability of the base and pavement, and broken material keys together better than rounded material. Vibratory compaction can be required to increase the material's compressive strength by forcing it to key together.

ACP is a blend of asphalt cement and mineral aggregates that are usually no larger than 25 mm. Except for seal coats and special-purpose, thin topcoats, the minimum pavement course depth should be 40 mm or twice the maximum aggregate size, whichever is greater. The maximum aggregate size can be changed for various purposes; it can be made finer for a smooth surface or coarser for an open-graded surface that provides greater traction, better drainage, and hydroplaning control. A chip seal as the final lift also can provide an open-graded surface.

Pavements might be placed in multiple lifts, and they can each have separate specifications if the design calls for them; however, a thick ACP lift saves time and labor, is easily compacted, and can be placed in cold weather. Although it has been considered impractical to place an ACP lift that is less than 40 mm, except for seal coats and chip seals, recent installations of thin and very thin top courses with carefully selected materials have achieved satisfactory results.

A porous ACP can successfully suppress traffic noise but can become plugged with dirt from the traffic. As described in Chapter 4, most traffic noise originates at the tire-road contact point. The noise is a combination of impact, air compression in front of the tire, air expulsion behind the tire, and suction. French contractors, pushed by the public to reduce traffic noise, have developed thin and very thin top courses of porous ACP. They reduced the maximum aggregate size from 14 mm to between 6 mm and 10 mm and used amendments to the binder. The contractors use this method on suburban highways that have heavy traffic moving between 65km/h (40 m/h) and 115 km/h (70 m/h).

Some contractors construct a 40-mm first course of porous ACP (10/14 aggregate mix and few fines). The 20-mm top course has a highly cohesive binder, a maximum 6-mm aggregate, and 0.8 percent of glass fiber. Other contractors use a top course of porous ACP containing rubber crumbs and organic fibers. These mixes substantially reduce noise while maintaining a rough surface for traction.

The basic purposes of flexible pavement are the following:

To support wheel loads and distribute them to the base, subbase, and subgrade

To resist deformation of the pavement due to loading, starting, and stopping movements

To resist abrasion and loss of surface aggregate

To shed rainfall and limit water-penetration on the surface

To provide a smooth, skid-resistant surface.

PAVEMENT RECONSTRUCTION

The ACP is durable, but all asphalt mixes gradually lose volatile elements to evaporation and consequently become more brittle and subject to breakage. Regular inspections will determine a schedule for replacing lost surface volatiles or for adding a new layer. Pavement condition determines the renewal method:

1. A seal coat, which is a spray of hot asphalt, replaces the lost volatile elements and can renew pavement that has just begun to crack.

2. A chip seal, which is a seal coat with aggregate of 20 mm or less spread and rolled into it, is needed if the damage includes a loss of surface aggregate and a need for better traction.

3. A thin or thick layer of new ACP is added when there is more extensive damage, including alligator cracking, uneven settlement, and pavement breakage.

Extensive spot repair might be needed before the ACP layer if potholes or small areas of differential settlement have developed in the pavement. Once deterioration starts, ACP loses stability and cohesion rapidly. Early renewal clearly is easier and less expensive than late repair. A properly designed, constructed, maintained, and periodically repaved flexible pavement will last many years.

Potholes developing in properly constructed ACP almost always are caused by water intrusion into the subgrade. In cuts, potholes can be the result of an upwelling of ground water, and it must be properly drained before pothole repair can be successful. Most often, deepening the ditch or installing a drain under the ditch can lower the water table. In some cases, drilling to dewater a hillside can be necessary. If such springs are observed during construction, they should be dewatered in the construction project; however, some might develop later and require maintenance or inclusion in a rehabilitation project.

In fills, the water that causes potholes usually intrudes through pavement cracks formed when the asphalt's most volatile elements evaporate. Evaporation begins immediately and slows, more gradually in cold weather than in hot weather. As cracks begin to form in six to eight years, however, hot weather and the kneading of the pavement by traffic can seal the cracks temporarily. Initial water intrusion should be drained away on the subgrade or subbase surface, but excessive and continued water flow will cause voids under the pavement base and indentations where the pavement is under heavy wheel loads. Once cracks form, freeze-thaw action can widen, deepen, and spread them.

Pavement performance can be measured by the pavement's ability to carry the traffic load without excessive potholing, cracking, faulting, or raveling and its ability to provide a safe and comfortable ride. Safety is measured by the amount of friction between the tires and pavement, and comfort is measured by the pavement's smoothness.

The AASHTO method uses a Present Serviceability Index (PSI) in pavement inspection. An Initial Serviceability Index (ISI) is used for new pavements, and a Terminal Serviceability Index (TSI) determines the minimum serviceability at which rehabilitation is necessary.

PSI must be evaluated and reported in an ongoing field inspection program. PSI inspection schedules can be established based on pavement age and thickness, traffic volume, and climate. ISI is 5 for new pavements on all classes of roads, but different roads have the following TSI requirements:

3 or 2.5 for major highways

2 for collector highways

1.5 for low-volume roads with low budgets.

PSI evaluations, with their rate of change and approach to TSI, can be used to develop and justify a pavement program that includes schedules, estimates, and budgeting for seal coats, chip seals, and repaving. Because water penetration accelerates deterioration, pavement condition might be deteriorating rapidly at the TSI, and the goal should be to save money through timely action.

Pavement wear is due almost entirely to the repeated application of heavy truck axle loads. Automobiles, with axle loads nearly all in the 400- to 800-kg range, contribute little to the deterioration. In hot weather conditions, the way automobiles massage the flexible pavement can be a positive factor.

Vehicles up to and including two-axle trucks without duals are disregarded in pavement design, and others are converted to the equivalent of 8160 kg (18-kip) axle loads. A properly designed subbase and subgrade should survive repeated loadings and actually should increase in compaction and density under load. A minimum pavement thickness between 25 mm (surface treatment) and 100 mm and a minimum base thickness between 100 mm and 150 mm are recommended for Equivalent Single Axle Loadings (ESAL) of 8800 kg each. The total ESAL from the new road's opening to the time it reaches the TSI in the design year range from 50,000 to more than 7,000,000. Many engineers will exceed the minimum pavement and base thickness because of local conditions or the state's design practices. Minimum Depth ACP and Base is shown in Table 11.1.

RECYCLED ASPHALT

Weathered asphalt found in ACP that has completed its service life is still a valuable commodity, and so is the aggregate with it. It makes excellent fill, but it is more valuable when it is recycled in new paving. In-place, cold-mix recycling is the least expensive recycling

TABLE 11.1 Minimum Depth ACP and Base

ESAL	ACP (mm)	BASE (mm)
0 to 50,000	25*	100
50,001 to 100,000	38	100
100,001 to 150,000	50	100
150,001 to 500,000	65	100
500,001 to 2,000,000	80	150
2,000,001 to 7,000,000	90	150
Greater than 7,000,000	100	150

* Surface treatment, not ACP

method. The pavement is ripped, broken, pulverized, and mixed on-site. Admixtures in the mix are those necessary to renew the ACP strength and flexibility. They might be bituminous material, aggregate of required grade, or stabilizers. After recompaction, a new ACP wearing course is usually applied. The cold-mix procedure also can be done at a central plant where quality control and mix reliability are better, but the expense is greater because of the handling and hauling.

One alternative at the central plant is a hot-mix procedure, which mixes new asphalt with the recycled material. In this case, base material also can be included in the mix for a greater volume and more uniform quality of the recycled material. This is a favored option, especially in projects that include pavement widening.

ASPHALT CEMENTS AND SUPERPAVE MIXES

Asphalt is the residue that is left after the most volatile compounds have evaporated from petroleum. It can exist in nature or be a byproduct of liquid or gaseous fuels produced at refineries. It is black or dark brown, and it ranges from rubbery to brittle at ordinary temperatures. The various viscosities have many uses.

Medium-curing (MC) and rapid-curing (RC) cutback asphalts are used as aggregate binders. MC is composed of asphalt and slow-evaporating kerosene or light diesel oil. RC is composed of asphalt cement and gasoline or naphtha, and it is used in situations that require quick evaporation and an early semi-solid state. Emulsified asphalts have colloidal particles dispersed in water with emulsifying agents, such as tallow, gelatin, glue, or soaps of fatty and resinous acids. Asphalt is about two-thirds of the weight of the emulsion, which is easy to apply and requires no heat. Rapid (RS), medium (MS), and slow (SS) emulsion breaking times are available.

Asphalt cements that are used as binders in flexible highway pavement are made of hard asphalt that has been softened with nonvolatile oils and heated to a workable consistency. Asphalts used for paving are graded by their viscosity or by the distance a needle penetrates them in a standard test. Both tests are performed at a standard temperature.

The Strategic Highway Research Program (SHRP) developed the Superpave ACP mix-design system. With this system, the designer chooses combinations of binder, aggregate, and modifiers for the desired level of service performance under the project's specific traffic conditions. The specific conditions can include traffic loads, the desired pavement's structural section, the local environment, and required reliability. The aim is to create an optimal pavement system with an ideal blend of binder and aggregate to meet the project's needs at the lowest cost. Computer programs can help the designer make the choices.

The Superpave Performance-Graded (PG) binder system has binders to fit many local climates and conditions. For a highway mix, a PG 58 might be the choice. Traffic that stops and starts has a tendency to shove a pavement made for highway use, especially in hot weather. Under these lateral forces, an asphalt-rich binder can lubricate the aggregate instead of binding it. A warmer, stiffer binder such as PG 70 should be used at intersections. If the PG series is not available, a viscous asphalt should be chosen, and the mix should be lean of asphalt. The Asphalt Institute, AASHTO, and ASTM have specifications for asphalt binders.

Superpave tends toward harsher mixes containing larger, more economical rock. Manufacturers are beginning to offer equipment that has been modified to handle Superpave mixes better, which is an indication of the innovation's success.

ADDITIVES AND RUBBER CRUMBS

The 13th annual Specifier's Guide to Asphalt Modifiers was compiled by *Roads and Bridges* magazine and was published in the May 2000 issue. The guide lists 63 commercial suppliers of modifiers that can be added to asphalt hot mixes. Some of the suppliers offer more than one product. The general categories of the additives are thermoplastic polymers, adhesion promoters, aging inhibitors, fillers, reinforcing agents, extenders, and emulsifiers. In addition to the actions inherent in these names, some of these additives can be used to improve water resistance, to prevent drain-down in coarse-graded aggregate mixes, and to provide resistance to rutting, shoving, crack formation, or potholing. This range of additives provides a fantastic toolbox for the flexible pavement designer.

Sulfur-asphalt mixes are used where sulfur, an oil-refinery byproduct, is less expensive than asphalt. They have been used in these locations for many years; in recent years, they have had the added advantage of reducing the import of petroleum products, a major component of our unfavorable trade balance. The sulfur stabilizes the mix and increases its compressive strength and its bond with aggregate particles. About 20 percent by weight of liquid sulfur will dissolve in hot asphalt, and an additional amount, up to 30 percent, can be dispersed in the mix. The addition of up to two parts per million of silicone to hot asphalt reduces pulling and tearing at the screed, and it increases moisture release. In the sulfur-asphalt mix, silicone also stabilizes the emulsion and improves workability.

The melting point of sulfur is 238°F, and its best working range of 270°F to 290°F is compatible with the 255°F to 300°F working range of asphalt. Sulfur-extended asphalt (SEA) requires little modification of the hot-mix plant or of normal paving procedures. Sulfur dioxide, an irritant, and hydrogen sulfide, a foul odor, are produced at and above 300°F and can be released at the plant or at the paving machine. Inflammability of both sulfur and asphalt increases rapidly above 300°F. The mix should not be heated above

295°F. All mix should be placed when it is at 290°F or below, and it should not be allowed to cool below 260°F.

In permafrost areas, 2 m of gravel base is required under pavement. This can be reduced to an 80-mm to 100-mm insulating layer of sulfur foam with 0.9 m of gravel cover.

Hydrated lime, which is used as fine filler in sand mixes, is a chemical additive in asphalt concrete. It makes the mix water-resistant, harder, stronger, and more stable. This enables faster compaction and gives the asphalt concrete greater density.

Recycling old tires produces rubber crumbs. When these are used in ACP, they reduce raveling and voids, lessen ACP's susceptibility to temperature variations, and promote aggregate adhesion. The crumbs can be added to the top course of new construction or to a wear course on old pavement. They can be mixed into ACP or spread before compaction.

STONE MATRIX ASPHALT

Another European innovation in asphalt-paving technology is Stone Matrix Asphalt (SMA). It was developed in Germany around 1965, and it is now in use throughout Europe. It has been studied in the United States since the late 1980s, and several states have demonstration projects.

SMA aggregate is gap-graded, has a high percentage of sizes between 5 mm and 12 mm, and is cubically broken for improved durability and strength. The voids in the coarse aggregate are filled with a sand-asphalt mix containing a fine filler, and the mix is stabilized with cellulose fiber.

When producing SMA pavement, carefully following guidelines is essential. Temperature guidelines must be observed carefully during production and placement, attention must be given to the cleanliness of the trucks and to the compaction of the mix in the paver, and rolling must be done vigorously and immediately behind the paver.

SMA pavement has excellent traction, is resistant to wear and weathering, and is suitable for hot, cold, wet, or dry climates. A comparison of SMA aggregate with a commonly used American aggregate for ACP is shown in Table 11.2.

TABLE 11.2 SMA Aggregate-American Aggregate Comparison

Percentage retained on sieve (MM)	AB 0/115	SMA 0/115
16	0	0
11	2	2
8	18	36
5	17	22
2	20	14
0.71	13	6
0.25	18	8
0.09	4	2
Passing 0.09	8	10

CHAPTER 12
AIRPORT ENGINEERING

AIRPORT ACCESS

Despite growing airport and airway congestion, we are not building many major airports. In nearly half a century, only two major airports—those at Dallas and Denver—have been constructed. Most airport layout work involves trying to make things work well. Although air transport is a mature industry, and its engineers have had time to solve problems and design efficient facilities, work still must be done to improve major airports and to build smaller ones. In major airports, at least until capacity is reached, safety and convenience are well served from the terminal entrance to the exit. As far as providing convenient access between home or office and the terminal, the American air transport industry, however, needs improvement.

We know that the most efficient way to feed a mass transit mode is to use another. Where would Heathrow be without London transit? In the United States, however, few rail transit systems feed major airports. Airports are way out there, building tracks that far away would be expensive, and anyway, it's someone else's job to plan and finance transit. So we depend on autos, which means hectares of parking lots, long walking distances, shuttle buses, and kiss-and-ride congestion at the front door. Landside congestion is bad and getting worse. Congestion at the passenger-drop-off area is a nightmare, and those picking up passengers frequently must park and walk in.

Some airports do better than others, but too frequently, traveling to the airport is a monumental waste of passenger time. If a trip includes hours between leaving home and becoming airborne, slower modes of transportation look better than air travel. Of course, the lost time is regained once the plane is in the air, and that's why the airlines are still in business. But why lose so much time when it is so valuable? We need to revisit some basic principles. When we subsidize something that doesn't solve the problem, we get more of the problem. We build more satellite parking and need still more.

- What might work?

    Talk to transit people about cooperative finance. Shared cost might be less expensive than failed solutions.

    Run shuttle buses with coordinated schedules to the modes of transit.

    Conduct demographic studies for bus routes and find out where airline customers are.

- Ask the passengers!

    Provide on-call bus service to airline customers' residences or office buildings.

    Set up bus collection stations and provide well-publicized schedules.

We got into this, as good engineers sometimes do, by providing good solutions to only one problem at a time. We need to look at the whole picture. Any airport-enhancement project should improve access, and airport access begins at the customer's premises. Clearly, an engineering and economic analysis should result in major-airport access that is better than the chaotic, costly, and time-consuming facilities that are too often prevalent.

RUNWAY CONFIGURATIONS

The simplest airport layout consists of a runway and a terminal area with an adjacent taxiway between them. This layout is adequate for a total of 50 aircraft takeoff and landing movements per hour. When traffic volume exceeds 50 movements per hour, the lowest-cost option is to construct a parallel runway beside the first one and taxiways between the runways. One runway can be used for landings and the other for takeoffs, but not simultaneously. The capacity in this configuration can be 70 movements per hour. In an area of strong and changeable winds, a "V" configuration can help air traffic avoid the strong crosswinds. In this case, the volume can be 50 movements per hour for one runway and 70 movements per hour for two. The angle between the two runways can be as high as 90°, depending upon the direction of the prevailing strong winds.

If there will be more than 70 movements, separating the two parallel runways by 1529 m (5000 ft) will permit simultaneous operation and serve up to 100 movements per hour. This leaves room between the runways for taxiways, the terminal, aprons for loading and unloading passengers and refueling and servicing aircraft, road access to the terminal, and hangars for repairs. With this configuration, a future parallel runway with taxiways can be added on each side. Added capacity would be 20 movements per hour for each runway, for a total of 140 movements per hour.

RUNWAY GRADES AND SLOPES

An airport site should be well drained, have well-graded sandy gravel soil, and be relatively flat longitudinally. According to FAA regulations (Part 25), a runway must be long enough for a plane to accelerate to the point of takeoff and, in case of a critical engine failure, for the plane to brake and stop within the runway limits. When landing, the plane should clear the approach-end of the runway by 50 ft (15.3 m) and be able to stop within 60 percent of the runway length.

Longitudinal grades should not exceed 2 percent for airplanes traveling slower than 120 knots or 0.8 percent for planes traveling faster than 120 knots. Grade changes should be avoided; if they are unavoidable, however, grade changes and their vertical curves should not exceed the values shown in Table 12.1.

TABLE 12.1 Grades and Grade Changes

Aircraft Approach Category (knots)	A to 90	B 91-120	C 121-140	D 140-165
Maximum Transverse Grade (percentage)	2	2	1.5	1.5
Maximum End Grade	0-2	0-2	0-0.8	0-0.8
Maximum Middle Grade	0-2	0-2	0-1.5	0-1.5
Maximum Grade Change	2	2	1.5	1.5
Minimum Vertical Curve Length	90 m	90 m	300 m	300 m
Minimum Distance Between Vertical PIs	90 m (A+B)	90 m (A+B)	300 m (A+B)	300 m (A+B)

Note: (A+B) is the total percentage of the two grade changes.

A runway's length depends upon the wingspan and approach speed of the largest and fastest aircraft it is built to serve. Runway length can vary from 2800 ft (855.9 m) for small aircraft with approach speed slower than 90 knots to 12,000 ft (3670 m) for aircraft with 261-ft (79.7-m) wingspan and an approach speed of 165 knots. The runway width will vary from 75 ft (22.9 m) to 200 ft (61.1 m).

A graded safety area that is 240 ft (73.3 m) to 1000 ft (305.8 m) long and has a downward slope of 0 to 3 percent is required at each runway end. The safety area will vary in width between 120 ft (36.7 m) and 500 ft (152.9 m). The values in Table 12.2 are based upon 1000-ft (305.8-m) elevation above mean sea level and a mean daily maximum temperature of 85° F in the hottest month. Runways for airplanes greater than 60,000 lb might require more length to accommodate heavy fuel loads; all pilots know they should be wary of heavy loads when temperatures soar or barometric pressures fall.

Unpaved shoulders can be as steep as 5 percent in the first 3 m adjoining the pavement. After the first 3 m, the shoulder for category C and D aircraft (those with approach speeds up to 165 knots) should not exceed 3 percent. Shoulders should be 38 mm below the pavement edge to prevent turf growth that holds water on the pavement. Subsurface and surface drainage on runway shoulders is similar to the drainage of highway pavements. See Chapter 5, Roadway Design.

TABLE 12.2 Runway and Safety Area Lengths and Widths

Approach Category A & B, Design Group	Runway Length m	Safety Length m	Runway Width m	Safety Width m
Minimal Wingspan	853	183	23	91
I, Wingspan to 15 m	975	193	30	91
II, Wingspan 15-24 m	1332	183	30	91
III, Wingspan 24-36 m	1634	244	30	122
IV, Wingspan 36-52 m	1942	305	46	152
Approach Category C & D, Design Group				
I	1673	305	30	152
II	1942	305	30	152
III	2222	305	30	152
IV	2920	305	46	152
V Wingspan 52-65 m	3261	305	46	152
VI Wingspan 65-80 m	3658	305	61	156

FLEXIBLE PAVEMENTS

Pavements can be either flexible or rigid. Subgrades must be composed of superior material and be thoroughly compacted so they are dense enough to support heavy loads. Subbase is required unless the subgrade is of the highest quality. An open-graded base layer of hard crushed stone provides drainage under the pavement. A gravity drainpipe gathers the water and transports it at a minimum grade of 0.5 percent to a disposal area. The FAA may approve a one-third reduction in the base course thickness when high quality local aggregates are used and are stabilized with asphalt or portland cement treatment.

Areas with the heaviest loading, called critical areas, will require the thickest pavement. These critical areas are the runways, and taxiways within 200 ft (61.1 m) of an intersection. The pavement in critical areas should be at least 100 mm thick. Non-critical areas can be 70 percent of critical-area thickness, but they should be no thinner than 75 mm.

Table 12.3 is entered with the California Bearing Ratio (CBR) for the base and subbase. The ACP thickness required for the necessary total strength is read under each increment of the aircraft gross weight. These readings can be used without adjustment for the number of annual departures. Table 12.4 for single-wheel landing gear to 75,000 lb, Table 12.5 for dual-wheel gear to 200,000 lb, and Table 12.6 for dual-tandem gear to 400,000 lb are read similarly, but all readings from these three tables are for 1200 or fewer annual departures. Each table shows the percentage of increase to apply to these readings for annual departures of 3000, 6000, 15,000, and 25,000. The pavement thickness shown in Table 12.3, Table 12.4, Table 12.5, and Table 12.6 is critical-area thickness for runway and intersection use.

TABLE 12.3 Flexible Pavement Thickness (m) for Light Aircraft

CBR	Gross Weight at 1814 kg (4000 lb) to 13,600 kg (30,000 lb)						
	1814	3628	5443	7257	9072	10,836	13,608
4	0.330	0.330	0.356	0.381	0.406	0.457	0.508
5	0.279	0.305	0.330	0.356	0.356	0.406	0.457
6	0.254	0.456	0.279	0.305	0.330	0.381	0.406
7	0.229	0.254	0.254	0.279	0.305	0.339	0.361
8	0.203	0.229	0.229	0.254	0.279	0.305	0.356
9	0.178	0.203	0.229	0.220	0.254	0.279	0.330
10	0.152	0.178	0.203	0.229	0.254	0.254	0.305
11	0.152	0.178	0.178	0.203	0.229	0.254	0.279
12	0.152	0.152	0.178	0.203	0.203	0.229	0.254
13	0.129	0.152	0.152	0.178	0.203	0.229	0.254
14	0.127	0.152	0.152	0.178	0.203	0.229	0.254
15.	0.127	0.127	0.152	0.152	0.178	0.203	0.224
16	0.127	0.127	0.152	0.152	0.178	0.203	0.229
17			0.127	0.152	0.178	0.178	0.229
18			0.127	0.152	0.152	0.178	0.203
19				0.127	0.152	0.178	0.203
20				0.127	0.152	0.178	0.203
25				0.127	0.152	0.152	0.178

Runways and taxiways for very small private aircraft, or for very occasional use, can be of well-maintained and well-drained turf or gravel. For safety, spacing between aircraft takeoffs and landings on these airports must be considerably longer than on paved runways, and following distances between taxiing aircraft must be greater to avoid gravel displaced by propeller action. Turf facilities should not be used when softened by rainfall, and gravel should be of sufficient depth to provide all-weather use.

When traffic increases to make these restrictions an operational problem, the runway or taxiway, or both, should be paved. If the problem is only with flying gravel, a light bituminous surface may suffice. Otherwise, the paving should be in accordance with Table 12.3.

TABLE 12.4 Flexible Pavement Thickness (m) for Single-Wheel Landing Gear

1200 ANNUAL DEPARTURES

	Gross Weight 13,608 kg (30,000 lb) to 34,020kg (75,000 lb)			
CBR	13,608	20,412	27,216	34,020
3	0.508	0.660	0.711	0.838
4	0.432	0.539	0.660	0.711
5	0.381	0.457	0.539	0.660
6	0.356	0.432	0.508	0.584
7	0.330	0.406	0.457	0.533
8	0.305	0.356	0.432	0.483
9	0.279	0.330	0.406	0.457
10	0.254	0.305	0.381	0.432
15	0.178	0.229	0.305	0.330
20	0.152	0.178	0.229	0.254
25	0.127	0.152	0.203	0.229
30	0.102	0.127	0.152	0.178
40	0.102	0.102	0.127	0.152
50	0.076	0.102	0.102	0.127

To convert tabular value of 1200 departures, multiply by 1.08 for 3000 departures, 1.16 for 6000 departures, 1.24 for 15,000 departures, and 1.32 for 25,000 departures.

TABLE 12.5 Flexible Pavement Thickness (m) for Dual-Wheel Landing Gear

1200 ANNUAL DEPARTURES

	Gross Weight 22,680 kg (50,000 lb) to 107,200 kg (200,000 lb)				
CBR	22,680	34,020	45,360	68,040	107,200
3	0.635	0.813	0.991	1.219	1.346
4	0.533	0.686	0.813	0.991	1.168
5	0.457	0.584	0.711	0.838	1.016
6	0.406	0.533	0.635	0.762	0.889
7	0.381	0.457	0.559	0.686	0.787
8	0.330	0.432	0.508	0.610	0.737
9	0.305	0.406	0.457	0.559	0.686
10	0.254	0.356	0.432	0.533	0.610
15	0.203	0.303	0.356	0.432	0.483
20	0.152	0.229	0.254	0.339	0.406
30	0.102	0.152	0.178	0.229	0.279
40	0.076	0.102	0.152	0.178	0.229
50	0.076	0.102	0.127	0.152	0.178

To convert a tabular value of 1200 departures, multiply by 1.04 for 3000 departures, 1.08 for 6000 departures, 1.12 for 15,000 departures, and 1.16 for 25,000 departures.

TABLE 12.6 Flexible Pavement Thickness (m) for Dual-Tandem Landing Gear

1200 ANNUAL DEPARTURES

	Gross weight 43,360 kg (100,000 lb) to 181,440 kg (400,000 lb)				
CBR	43,360	68,040	107,200	136,080	181,440
3	0.711	0.914	1.092	1.372	1.600
4	0.610	0.762	0.925	1.168	1.397
5	0.533	0.686	0.813	0.991	1.219
6	0.457	0.610	0.737	0.864	1.067
7	0.406	0.533	0.660	0.787	0.940
8	0.356	0.483	0.584	0.711	0.838
9	0.330	0.432	0.533	0.660	0.762
10	0.305	0.381	0.508	0.610	0.711
15	0.229	0.330	0.381	0.457	0.584
20	0.178	0.254	0.305	0.3812	0.432
30	0.127	0.178	0.203	0.254	0.330
40	0.102	0.127	0.152	0.203	0.229
50	0.076	0.102	0.127	0.178	0.203

To convert a tabular value of 1200 departures, multiply by 1.04 for 3000 departures, 1.08 for 6000 departures, 1.12 for 15,000 departures, and 1.16 for 25,000 departures.

A binder course of ACP with a maximum aggregate size of 50 mm or half the course thickness is placed on the base and covered with a wearing course that has a maximum aggregate size of 10 mm to 25 mm. The usual maximum size is 20 mm.

For example, an airport serving aircraft with dual-tandem landing gear and gross weights up to 300,000 lb is projected to serve 25,000 departures in its 20th year, which is the design year. The following calculations find an appropriate base course and asphalt concrete pavement:

According to Table 12.6, interpolating between CBR of 10 and 15 at 300,000 lb gross weight, CBR 14 yields 0.457 m + ⅕(0.610 − 0.457) m = 0.488 m.

Adjusting to 25,000 departures, 0.488m (1.16) = 0.566 m total thickness of base and ACP.

With a 0.05-m ACP wearing course and a 0.15-m ACP binder course, the remaining 0.366 m is the required base. But if the FAA approves the material and the stabilization of the base, it can be reduced: 0.366 m (⅔) = 0.244 m will be sufficient.

RIGID PAVEMENTS

For small aircraft with a gross weight under 12,500 lb, the minimum portland cement concrete (PCC) pavement thickness is 125 mm. For planes with gross weight up to 30,000 lb, the minimum thickness is 150 mm. For aircraft with gross weight between 30,000 and

400,000 lb, the pavement thickness is determined by the following on FAA "Airport Improvement" charts with the entering data:

1. Foundation modulus k of the subgrade modified by the subbase, in psi
2. Flexural strength of the concrete, about 10 percent of compressive strength, in psi
3. Gross weight of the heaviest aircraft the airport plans to serve, in pounds
4. Annual departures expected in the design year, usually the 20th.

PCC reinforcement and joint arrangement is similar for airport pavements and highway pavements, except that some airport pavements are much thicker and require larger reinforcing bars. Reinforced pavement is recommended, and rebars vary in diameter from 19 mm for 0.15-m pavement to 51 mm for 0.61-m pavement. Prestressed PCC pavement can yield substantial savings when it is compared with continuously reinforced pavement; the thickness might be halved, and the amount of steel required might be reduced by four-fifths. See Chapter 10, Rigid Pavement Design.

Expansion joints are necessary near buildings and where runways and taxiways intersect. The reinforcing dowels at the expansion joint hold slab alignment and bear the load of the aircraft. They should be securely fastened into one pavement and be free to move with the expansion in the other, preferably alternating, pavement. If dowels cannot be used, a thickened edge should be formed and poured. Transverse contraction joints are formed or sawed across unreinforced concrete slabs at 4.5-m to 7.5-m intervals.

CHAPTER 13
RETAINING WALLS

CHOOSING RETAINING WALLS

A retaining wall can be a design solution for one or more of the following reasons:

The high urban right-of-way cost. This can make the cost of space for cuts' back slopes comparable to the cost of retaining walls.

Public resistance to taking property in urban areas. The need to depress the roadway for noise control can exist concurrently.

Unstable material, such as sand that flows in dry and windy conditions or a clay layer that expands when it is wet and creates a slip plane in a cut slope. This unstable material might endanger the completed roadway or buildings beside it as a result of roadway excavation.

The necessity of locating a road on a steep side slope.

Retaining-wall types include the following:

Walls made of mechanically stabilized backfill and reinforced earth that use in-place material

Gravity walls that depend upon their mass to resist lateral earth forces

Cantilever walls that resist lateral earth forces through cantilever action of a vertical stem wall with a horizontal base

Counterfort walls that have a base extended from the wall to support counterforts (wedge-shaped cantilevers) within the backfill.

The choice of a wall type is governed by the required height, the relative costs of the different types, and the construction difficulties experienced in the unique field conditions at each site. For all wall types, an additional safety factor can be built-in with earth anchors or anchors that are grouted into rock. Designs and specifications should follow AASHTO Standard Specifications for Highway Bridges. For design information that goes beyond the discussion in this chapter and is especially for walls higher than 3 m, see Structural Engineering, another book in this series.

If the wall is of variable height, a 1-m section at the highest point should be analyzed. Moments are taken about the toe, including, in appropriate climates, up to 5000 kg/m for frost action at the top of the wall, disregarding the weight of soil above the toe. The sum of the overturning moments should be no more than 75 percent of the righting moments.

MECHANICALLY STABILIZED BACKFILL AND REINFORCED EARTH

Backfill can be stabilized in several ways. When a fill slope of ordinary earth must be steep, a common and inexpensive way to stabilize it is to reinforce the edge of the fill with geosynthetics. Each successive fill layer is placed on one edge of the geosynthetic. The fabric then is folded up over the fill layer, and the next layer is placed on top of it. The slope should be about 0.25:1 or 0.5:1, depending upon the quality of the fill material. If a steeper slope is critical, metal facings with horizontal metal or fabric reinforcing straps in each layer of fill can be substituted for the geosynthetic at increased cost.

The geosynthetic is vulnerable to sunlight, and it should be covered by a more resistant material. Shotcrete is expensive and subject to cracking. Asphalt sprays have been successful, as have facing walls made of various materials, such as stone, tires, concrete panels, and treated timber.

An unreinforced mass concrete wall might be a solution to earth stabilization in some cases, such as a rocky cut slope with some loose rock. The wall's slope would be the angle of repose of most of the material, and the wall would key into the rock slope and should be stable. This would be an expensive way to control falling rock, but other solutions might be even more expensive.

The Reinforced Earth Company has developed a patented standard wall. It offers several choices of concrete or metal facings bolted to thin metal or geosynthetic strips in the backfill that will develop the frictional resistance required to support the facing. The desired architectural treatment would determine the facing choice. Keystone has patented a block wall with modules that fit together and have several architectural appearances. The backfill is reinforced with geogrids.

GRAVITY WALLS

Gravity walls usually will be the type that is chosen up to 3 m high; those between 3 m and 5 m might compete economically with cantilever walls, depending upon local conditions. They can be constructed of masonry up to 1.5 m in height, increasing in thickness from base to top. The base should have extensions for a toe that will resist overturning and for a heel that will be located under the compacted backfill. Masonry gravity walls should adhere to the highest construction standards because they will be in continuous lateral stress. If the backfill is subject to water accumulation, adequate drainage should be installed, with gravel backfill immediately behind the wall and appropriate wall waterproofing.

Proprietary bin walls made of circular or rectangular steel can be used up to heights of about 1.5 m. The allowable height can be greater when the walls are used inside curves, when earth anchors support them, and when their use on both sides of the road permits the use of cross ties.

Closed-face bins made of stacked steel bins filled with earth and rock make an effective and economical retaining wall. They can be stacked vertically or battered. When they are battered, they can be used up to heights of 9 m. The steel bins and similar crib walls made of treated timber or concrete beams are built like log cabins, and they rely on the stabilized fill inside the crib for stability. They take more space and more backfill work than other designs, but they seldom require earth-anchor support. The open face of the crib sometimes allows some of the material to run out, and even treated timbers are impermanent when they are buried in a fill. The concrete beam cribs are the most expensive but are more permanent, and they can be built with a smooth face on the outside to reduce material loss.

Gabion walls are made of preformed wire baskets filled with free-draining rock. Their use can be limited if this material is not readily available. Little foundation preparation is required, because gabion walls can survive considerable differential settlement with little distress. Gabions can be stacked to heights of 9 m; however, if they are to be higher than 7.5 m, the foundation conditions should receive more attention, and they should be observed closely throughout construction for wire damage and differential settlement.

Monolithic PCC can be used for retaining walls that are up to 5 m high. They can be faced with masonry or decorative materials for aesthetic reasons. The cross-sectional shape of a PCC gravity wall is a right triangle set on a rectangular base. The hypotenuse should face toward the backfill and the top angle should be attenuated. The top of the wall can be made as wide as desired to increase the mass until the shape becomes a rectangle rather than a triangle. The base should extend beyond the wall for a toe and a heel, each at least 0.3 m, and the base should extend the full length of the wall (Figure 13.1).

Compressibility under the toe of the base of the gravity wall is particularly critical, because subsidence under the heaviest wall loading can cause overturning. For compressible soils, soil amendments, such as gravel, lime or pozzolans, should be considered, and fill compaction should be done carefully. If the soil quality is in doubt, the base should be enlarged to bear on a greater soil area.

Contraction joints should be a minimum of 10 m apart for crack control, and expansion joints should be a minimum of 30 m apart.

The vertical resultant Rv should be located in the middle third of the base to prevent tension in the wall. The distance from the toe of the base is the sum of the moments about the toe divided by Rv. To determine whether allowable soil pressures are exceeded, the range of pressures on the soil is found with the following formulas:

$$P_{max} = Rv/A(l + 6e/L), \text{ and}$$

$$P_{min} = Rv/A(1 - 6e/L)$$

where A = Area of the base
 L = Width of the base
 e = Distance, parallel to L, from the centroid of the base to Rv.

CANTILEVERED AND COUNTERFORT WALLS

A sheet-pile wall and an H-beam pile wall with lagging are cantilevered walls. The piles should be driven deep enough to assure stability, which means driving them to bedrock if this is feasible. If the bedrock is too shallow and drilling holes for the soldier beams is impractical, the sheet pile or beams can be cast into a concrete foundation keyed into the

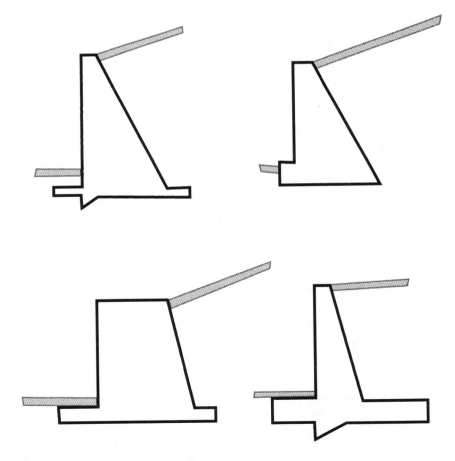

FIGURE 13.1 Monolithic PCC gravity walls.

bedrock. Unless the construction is in an area that is so remote it would greatly increase pile driver and crane costs, the most economical solution to building retaining walls on steep slopes is likely the sheet-pile wall or the H-beam wall. Little excavation is required. Down-slope environmental damage should be little or none.

When they are driven, sheet-pile walls are complete and ready for soil anchors and backfill. Anchors or soldier piles can reduce the driving depth and the number of the costly sheets that are needed. An H-beam wall requires fewer piles to be driven, but it requires lagging before backfilling can begin. Anchors or soldier piles between the piles holding the lagging are less likely to be needed. Depending upon the earth forces involved, timber, precast concrete panels, or horizontal sheet piling can be used for lagging.

Cantilevered walls of reinforced PCC can be economically competitive with gravity walls at heights of 3 m to 4.5 m; at heights of 4.5 m to 6 m, cantilevered walls can compete economically with counterfort walls. They present a vertical face to the roadway, and they have a top as thin as is convenient for concrete pouring: 0.2 m or 0.3 m. The thickness of the stem at the base is at least double the thickness at the top. The base is poured with a key to inhibit sliding. A depression above accommodates a keyed pour of the stem (Figure 13.2).

The stem is designed to overcome the bending moment and shear due to the lateral earth pressure on the wall. The base is designed to prevent sliding and overturning and to remain within tolerable pressures on the soil underneath. Main vertical reinforcement is placed parallel to the back face of the stem with 75 mm of concrete cover, and it is tied to reinforcement extended from the base and to horizontal steel of the same size.

The AASHTO Standard Specifications for Highway Bridges require at least 0.05 cm^3 of reinforcing per meter of wall height to resist horizontal and shrinkage stresses. The smaller steel is placed vertically and horizontally in the front face of the stem wall and horizontally above the main reinforcement. The base is reinforced with sizes and spacing as required in the bottom of the toe and the bottom of the heel with 75-mm coverage. The base steel can be extended into the key if desired. The base is poured first, leaving vertical bars exposed to dowel the stem wall and the counterforts to the base.

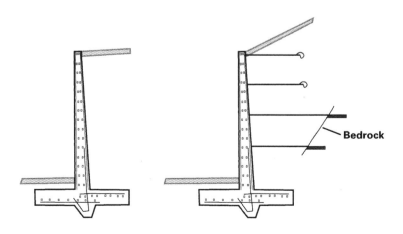

FIGURE 13.2 Cantilevered PCC walls.

Counterfort walls are appropriate when they are taller than 6 m, and sometimes when they are shorter than 6 m. Counterforts are triangle-shaped tie walls that are perpendicular to the stem wall and buried within the backfill. They are in tension between the base and the stem. Buttressed walls are similar supports on the compression side, but they are not useful in transportation because they would take up roadway space and defeat the retaining wall's purpose (Figure 13.3).

Stability design analysis is the same for higher walls as for lower walls, except that the design section should be from center to center of the counterforts rather than a 1-m section. The two faces of the stem wall are parallel, and the wall thickness usually is 0.3 m or more. The base is poured first, leaving vertical bars exposed to dowel the stem wall and counterforts to the base.

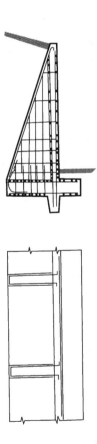

FIGURE 13.3 Counterfort PCC wall.

CHAPTER 14
HIGHWAY-RAILROAD CROSSINGS

RURAL CROSSINGS

Generations were taught to "stop, look, and listen" at remote railroad crossings, but because of familiarity (and sometimes because of unfamiliarity) we have lost some of this awareness. Rural road railroad-grade crossings are serious safety hazards. Although most such crossings have had few or no accidents, they are more hazardous per crossing vehicle than are crossings that have a higher usage and more warnings and traffic controls. Minimal warnings at these remote crossings should be "Stop" signs preceded by "Stop Ahead" signs.

Where "Stop" signs are not installed, drivers must be warned to slow down and be prepared to stop. Drivers must be able to determine whether a train is in a position that requires them to slow or stop. Where possible, anything that obstructs the driver's view of approaching trains should be removed. Depending on the types of obstructions that remain, speed limits must be set that assure the driver's ability to stop when an approaching train is observed.

Where feasible, lightly used rural crossings should be closed, the crossing road destroyed, and the crossing needs served at a combined facility. It is more feasible economically to provide such combined crossings, which have better safety devices and adequate sight distances. Road system improvement in recent decades has provided opportunities to combine rural crossings, but not all of these opportunities have been recognized. Civil engineers responsible for rural road safety should be alert for these situations, even those involving private roads. Engineers should have the full support and gratitude of railroad management when taking a corrective action.

Railroad grade crossings should be as flat as possible: They should be in the same plane as the rail tops for 0.6 m outside each rail. Beyond that distance and within 9 m of the rail, the road elevation should be not more than 75 mm higher or 150 mm lower than the top of the nearest rail. Unless the railway is in superelevation, this will mean the gradient should be between 1 percent and –2 percent. Many remote rural crossings will fail this test (Figure 14.1).

At-grade crossings on road or railroad curves should be avoided if this is practical, because of comfort difficulty at superelevations and because of the sight-distance safety problem. The above standards for a relatively level crossing cannot be met in a superelevated road curve. Uncomfortable grade changes are likely when a road crosses a superelevated railroad curve, and they are certain if both are in curve. In a road curve, the auto can be at an angle of less than 60° to the tracks; this requires the driver to look backward to see oncoming trains. Also, negotiating the curve can take the driver's attention away from watching for trains. If the railroad track curves, the driver's vision of approaching trains can be obscured.

All crossings should be as close to 90° as possible and never less than 60°. The roadway width should be continuous across the tracks, and the shoulders should be paved if there is pedestrian or bicycle traffic. Vehicles, bicycles, and pedestrians should cross the tracks without pausing; during a 10-second crossing time, a train traveling at 120 km/h covers 330 m (1100 ft).

Grade-crossing warnings are advance-warning signs, crossing signs, pavement markings, flashing-light signals, or signals with automatic gates. The American Association of State Highway and Transportation Officials (AASHTO) Manual on Uniform Traffic Control Devices (MUTCD) gives details about these devices. Selection criteria for warnings include crossing geometrics, the amount and speed of auto and train traffic, accident experience or risk evaluation, number of tracks, sight distance and terrain, and the volume of pedestrian and bicycle traffic. Train and auto traffic volumes are directly related to the economic feasibility of safety devices or traffic separations at the crossing.

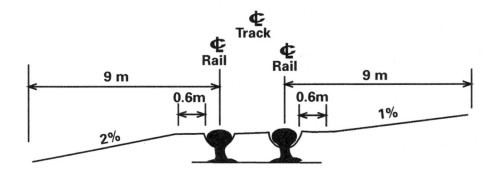

FIGURE 14.1 Road-railroad crossing.

Speed controls, especially in rural areas, are expensive for railroads and difficult to enforce for autos. The road planner must always be aware that railway engine operators are trained professionals, but they operate heavy vehicles with a long stopping distance. On the other hand, auto drivers vary in training and experience, but they operate vehicles that are much easier to control. The driver must be made aware of all circumstances affecting the need for vehicular control. The planner always must provide the best protection available under the circumstances.

In most cases, reasonable speeds should be assumed for the train and autos under the area's existing road conditions. Safety facilities, including the clearance of obstacles to provide adequate sight distances, should be planned accordingly. Accident records, near-miss narratives from local residents, and the proactive judgement of civil engineers are solid justification for safety expenditures.

Railroad crossings with more than one track are extreme hazards for auto traffic, because one train can obscure another from auto drivers' sight. Many drivers are unaware of this fact. Any grade crossing of multiple tracks should be evaluated for separation. If separation cannot be justified or if the crossing needs cannot be met at another separation, there should be, at minimum, an installation of automatic signals and crossing gates with advance warning signs and pavement markings. All multi-track crossings should be paved, even if traffic volumes do not justify paving the rest of the road.

Where a road is parallel to the tracks and a signaled intersection is adjacent to the railroad, the signal must have automatic railroad preemption, and "Do Not Stop On Tracks" signs must be installed.

When crossings do not require a stop and have no train-activated warnings, drivers must have enough sight distance to evaluate a safe time to pass through the crossing before the train's arrival. Also, the driver should have enough time to see the train to make a safe crossing or to come to a safe stop before entering the crossing area. See Table 14.1.

Construction should be considered to provide the necessary sight distance. This could include adding long vertical curves, widening ditches and flat back slopes in cuts, and removing vegetation. If sight restrictions remain, limiting vehicles' speed with road curvature, setting lower speed limits, and adding warning signs to inform the driver of danger might be necessary to assure a safe crossing. Design of the approaches to the crossing should not distract the driver's attention; they should allow the driver to fully concentrate on making the crossing safely.

For a driver stopped at the tracks to evaluate a crossing's safety, the required safe sight distance in meters is 4.53 times the train speed in km/h. This formula provides a safe distance, even if the train appears as the driver begins acceleration. The designer should assure this sight distance for the highest speed that trains reach at the location; for example, 453 m would be a safe sight distance for a vehicle crossing ahead of a train traveling at 100 km/h.

URBAN CROSSINGS

In urban areas, railroad crossings should be eliminated at minor streets, and the traffic should be redirected to the nearest arterial street. Another option is to combine crossings at one of the minor streets to reduce the number of crossings. Arterial crossings that are

TABLE 14.1 Required Design Sight Distance to Clear Single Track

[Train Distance From Crossing (m)]

Train Speed (km/h)	20 m Truck Speed (km/h)									
	30	40	50	60	70	80	90	100	110	120
40	80	75	75	77	81	85	89	94	101	105
50	100	94	93	96	101	106	111	118	126	132
60	120	112	112	115	121	128	133	141	151	158
70	140	131	131	134	141	149	155	165	176	183
80	161	151	151	154	162	170	177	189	202	211
90	181	168	168	173	182	191	199	212	227	237
100	201	187	187	192	202	213	221	236	252	264
110	221	206	206	211	222	234	244	259	277	290
120	241	225	225	230	242	255	266	283	302	326
130	261	243	243	249	263	275	288	306	327	343
140	281	262	262	269	283	298	310	330	353	369

only occupied by trains infrequently in off-peak traffic periods can be served reasonably with high-type gates and signals, and they are not usually separated. If arterial crossings are frequently closed to traffic or if some closures occur in peak traffic periods, the closure is a costly inconvenience. A separation should be provided as soon as funds are available.

All at-grade road-railroad crossings that are used by pedestrians or bicyclists should be paved, including the shoulders. Intersections of bicycle routes that are not perpendicular to the railroad require additional paving width so that the bicycle path can become a safe, perpendicular crossing. Sidewalks should be built at crossings to connect existing or planned sidewalks.

Floodlighting or highway lighting should be considered at crossings where trains are scheduled to run at night. If the area has no other lighting, however, a small lighting area that provides no gradual transition from dark to light can be a hazard because of the glare.

SEPARATIONS

Vertical clearance at underpasses should be 4.4 m or more. For freeways, arterials, and other roads that have occasional truck traffic with permitted high loads, the recommended clearance is 5 m. Clearance required at overpasses usually is 7 m, but some states require more. Road construction planning always should be coordinated with the railroad, and this is especially important in the early planning stage of an overpass. Double-stacked containers and other height requirements may apply. Because of the greater clearance required over railroads, the greater flexibility of road gradients, and the greater width of highway bridges, crossing under the railroad will usually be more economical.

Unless local conditions govern, bridge spans or undercrossing widths should include the road shoulder and any additional lanes to be built in a 20- or 30-year horizon. Urban road-railroad separations should always provide safe facilities for bicyclists and pedestrians.

When an arterial that is parallel to the tracks intersects an arterial that is crossing the tracks, a separation over both the railroad and the parallel arterial might be warranted. If there is no separation, the intersection signal should be interconnected with an automatic, railroad-controlled crossing signal, and signs should warn traffic not to stop on the tracks.

CHAPTER 15
BICYCLE FACILITIES

BIKEWAYS

Bikeways or bicycle trails on separate rights of way are excellent recreational facilities. They provide exercise, usually in a park-like setting, and sometimes lead to a major attraction such as Mount Vernon. They are a valuable amenity of city living, promoting physical and mental health. A bikeway can be established on new right of way or on abandoned rail or canal right of way. Unlike motorized vehicles, bicycles are not unwelcome in pedestrian areas and parks, but very few American bikeways are designed for commuters.

In 1970, the mayor and city council of Eugene, Oregon, established a bicycle committee. Eugene, an industrial city with a university and a population of 150,000, had a 2-km bike trail in a city park. By 1982, the city had 16 km of bikeways and 60 km of bike lanes on streets. Bicycles were considered an essential part of traffic service and bicycle facilities had an annual budget. Davis, California, and a few other American cities have had similar experiences, as have many European cities and the whole nations of Denmark and the Netherlands. Most American cities have more bicycles than automobiles, but most of the bicycles are in storage because of inadequate facilities. The bicycle, invented shortly before the automobile, is the most efficient personal vehicle ever devised for trips under 10 km, and we are ignoring its possibilities while cursing our traffic jams.

Bikeway grades should be moderate; sustained grades should be less than 5 percent. Grades shorter than 100 m can be up to 10 percent. Expert riders could consider them a challenge and an exercise in gear-changing; for others, this is not an unreasonable walking distance. If low traffic volumes are expected, the minimum width can be 1.2 m with frequent widened sections to accommodate meeting other cyclists, if sight distances are excellent. Widened sections should be continuous where sight distance is restricted.

In widened sections and when directional volumes are greater than 50 vehicles per hour, lane width should be at least 2.4 m and preferably 3 m to allow for a lane in each direction. When directional volume is 180 vph or more, a climbing lane should be added for upgrades greater than 3 percent.

For low traffic volumes, a well-graded and compacted gravel surface is acceptable. A light bituminous surface should be provided for directional volumes greater than 50 vph. The surface should be smooth, with a 15-mm maximum aggregate size.

BIKE LANES

Bike lanes run alongside automobile lanes on city streets. They are next to the curb, clearly delineated with a 100-mm white stripe, and are a minimum of 1 m wide. A width of 1.5 m is best. Care should be taken to correct hazardous conditions in this space, such as sewer intakes or grills that could trap a bicycle tire.

A speed limit that is safer for bicycles, such as 30 km/h, should be posted and strictly enforced, or a barrier curb should be installed to replace the white line and separate traffic. Traffic signs should identify the street as a bicycle route. Parking should be prohibited on bicycle streets because the parking cars and opened doors would force bicyclists into the hazardous traffic. Perhaps after a few years of education, such measures will be less necessary, but for now, autos and bikes must be considered incompatible. Bicyclists and drivers must be educated about the danger of left turns, or left turns must be prohibited for bicycles. Riders can become pedestrians to cross the two streets.

Transportation costs often prevent the poor from becoming employed. Even bus fare can be too much when the first paycheck is two weeks away. A bicycle could solve the problem, but it would not help an inexperienced rider in heavy, fast traffic.

In our gridlocked streets, bus riders frequently can make better time as pedestrians. When bike lanes are successful, bikes will pass the gridlock, and motorists will learn to leave their cars at home and participate in the transportation system that works. Bikes will not gridlock; even if cars block the bike lane, the bike can be walked around the obstruction.

BICYCLE STREETS

Humans are making the decisions, not the machines. We rightly reserved the freeways for ourselves as drivers. The time has come for us to decide what streets we will reserve for ourselves as bicyclists and as pedestrians. The political difficulty of this is well illustrated by the continuing controversy over the automobile restriction on one block of Pennsylvania Avenue that was requested by the U.S. Secret Service for the protection of our presidents. With that steering wheel in hand, we become a different person.

Only logic can save us, and who better to apply logic than engineers?

Let us imagine that a four-lane street is converted into a minimal service bicycle street; that is, bicycle service is a priority, but other things are changed as little as possible. Automobile parking would continue on both curb lanes, and automobile traffic would be discouraged although not prohibited. The speed limit would be set at 25 km/h, and automobiles would have to observe it. Drivers would choose less-restricted routes, using the bicycle street for short distances. If the slow speed limit did not dissuade automobile traffic, another means could be employed, such as prohibiting passing or prohibiting autos in occasional blocks. The street conversion would help relieve automobile congestion on parallel streets. See Table 15.1.

TABLE 15.1 Spacing and Lane Capacity of Bicycles

Speed Km/h	Spacing at 1.7 m/cycle m	Spaces	Bicycles / hour
4	5.1	3	1570
8	6.2	3.5	2670
12	6.8	4	3530
16	8.5	5	3760
20	10.2	6	3920
24	13.6	8	3530

To estimate capacity, assume 20 km/h is the average speed, there are five empty spaces behind each 1.7-m bike, and there are two 1.8-m bike lanes in each direction.

$$(2)(20,000 \text{ m}/6)(1.7 \text{ m}) = 3920 \text{ directional vehicles per hour.}$$

Assume a 50-percent reduction because of traffic signals and other obstructions, and the capacity is 1960 vehicles per hour.

By comparison, a 3.6-m lane with these same conditions would carry about 800 automobiles per hour, nearly all of them with only a driver, and a free-flowing arterial or freeway lane with traffic separations would carry about 2000. The conclusion is that a minimal bike street doubles capacity. Put another way, the capacity goal for a 8000-DDHV urban freeway can be reached by turning four surface streets to minimal bike streets. With persuasion and safety, people will bring their bikes out of storage.

Another plan could be to convert a four-lane street to a maximum service bicycle street where automobiles are prohibited and the cross traffic is eliminated except at separations. Four bike lanes would be available in each direction for normal traffic, while commuter traffic would require six bike lanes in the principal direction. The latter would triple directional capacity to 11,760. Of course, some riders would fall and others would lose bike chains, but it appears that an ordinary four-lane street converted to a maximum service bicycle street would approach eight-lane freeway capacity. With reversible lanes, it would exceed 10-lane freeway capacity.

URBAN CONGESTION SOLUTIONS

The convenience of private automobile transportation is often a reason people give when objecting to the use of mass transit. Never mind that we have already lost that convenience; who wants to cruise for half an hour hoping to find parking? There are two answers to this objection. First, our feet were made for walking, and most of us need the exercise. The second option, when walking is not practical, is the most efficient personal transportation machine ever invented: the exercise bike, returned to its rightful place in the street.

Transit trains and buses serve large city centers in a grid so dense that automobiles are not necessary. Only auto storage is necessary, and it should not be in the city center, but on the periphery. Automobile traffic should be permitted in specified arterials that are designed to provide through service. Trucks should make late-night deliveries for next-day use.

Dense city centers must be the domain of the pedestrian, who shares the area in part with transit and those using individual modes of transportation, such as bicycles, skates, skateboards, and wheelchairs. Automobiles have put our quality of life in disarray and have upset the mental balance of drivers. Road rage and impatience are the order of the day in city traffic.

Our cities should be more people-friendly. The need is clear, and so are the general outlines of a solution. The exact methods are far from clear, but they must be varied and local. The problems vary from Phoenix sprawl to Los Angeles freeway gridlock to Washington street gridlock. We have a complex web of open subsidies, special-treatment subsidies, and subsidies that come from passing costs to someone else. We get more of what we subsidize. If that is not what we want to happen, we must change subsidies, policies, and habits. When we face that squarely, we will be on the way to solving our urban problems, including congestion.

As has been demonstrated, if any portion of the local traffic problem is urban congestion, a major part of the answer is the inexpensive and easy but politically and socially difficult provision of adequate bicycle facilities. As the next chapter will show, transit can supply the rest of the answer. No one seems to have any other answer at any price.

CHAPTER 16
TRANSIT

According to the American Public Transportation Association, mass transit ridership rose 4.5 percent in 1999 to 9 billion trips, the most since 9.3 billion trips were made in 1960, before much of the Interstate Highway System had been built. Ridership and its increase in 1999 are broken down in Table 16.1.

TABLE 16.1 Mass Transit Ridership

MODE	1999 RIDERSHIP	INCREASE
Heavy rail (subway system)	2,686,000,000	6.47%
Trolley bus	126,500,000	6.14%
Demand response (elderly, disabled)	107,800,000	4.72%
Bus	5,360,400,000	3.84%
Commuter rail	393,700,000	3.76%
Light rail and other	383,500,000	0.91%

Transit use peaked at 23.4 billion trips in 1943 and declined to 6.5 billion in 1972. The numbers rose slightly for several years, and a steady increase began in 1995. Transit managers attribute the increase to traffic congestion, public spending, transit innovation, and a strong economy.

HOV LANES

How many passengers create high occupancy on High Occupancy Vehicle (HOV) lanes? Buses of all sorts automatically are considered HOV. Because they serve many more people per vehicle than automobiles, free-flow of buses is the first aim of HOV. When the HOV lane or lanes have excess room that other vehicles can use without interfering with the free-flow of bus traffic, vehicles occupied by car-poolers are allowed. Small, private buses not only qualify, they are usually encouraged by preferred or free parking furnished by employers. Whether passenger car HOV 2, HOV 3, or HOV 4 create high occupancy is the principal question to answer because HOV 5 is seldom required. The answer depends upon local conditions.

The aim is to encourage as much car-pooling as possible in the HOV lanes without approaching gridlock. The study must determine the highest rate at which car-pooling is attractive. Will HOV 4, three passengers, have less use than 67 percent of HOV 3, two passengers? Will HOV 3 have less than 50 percent of HOV 2? After the occupancy level with the most passengers is determined, the occupancy level of the HOV lanes is evaluated, including occupancies of passenger cars, buses, and emergency vehicles. The lanes must be free flowing to meet the objective, but if the occupancy level is noticeably low, a driver and a passenger in the gridlocked lanes will protest HOV 3.

An HOV lane can be the left or right lane of a multilane highway or arterial street. If it is the left lane of a freeway, HOV traffic must merge across the freeway to exit, and if it is the left lane of an arterial, non-HOV eligible vehicles must be permitted to enter the left-turn lane. If the right lane is designated for HOV, ineligible drivers must be permitted to exit and turn right, and HOV drivers on arterials must be able to merge left to make left turns.

HOV works best when the HOV lanes are on a separate roadway, usually a reversible expressway between the two directional roadways. One example is Virginia's Shirley Highway, Interstate 395 south of Washington, D.C. The ability to set barriers to wrong-way travel simplifies management, and enforcing HOV rules is easier. The Washington HOV has worked so well it spawned a culture of *slugs*, people waiting for rides from motorists who need them to enter the HOV lanes.

A well-planned HOV system is almost certain to be a success. Every day it demonstrates its value to thousands of prospects who sit and watch the express traffic go by.

BUS TRANSIT

Including bus service in HOV lanes is an important step in getting people to use mass transit. When it becomes apparent to the gridlocked rush-hour motorists that bus service takes less time, many who have work schedules too variable to carpool will be attracted to convenient bus service. The transit authority can facilitate this process by running buses with local pickup routes in the morning, having them turn express for the trip on the HOV lanes to the city center or another traffic generator, and reversing the routing in the afternoon.

A bus roadway that serves stations for passenger pickup should not be located in a freeway median because this would make passenger access difficult. Pedestrian bridges or tunnels might be required, and parking necessarily would be distant. A better arrangement is to have the bus roadway parallel to the freeway on one side, between the freeway and frontage road. A turnout from the frontage road at the bus pickup station would allow room for kiss-and-ride service, and off-street parking could be nearby. Bus slip ramps to and from the frontage road will permit collection, distribution, and transfers to express buses without interfering with freeway traffic. Rail transit could operate the same way without the slip ramps.

If a median is the only space available for a bus roadway, passenger stations can be built at separation structures between interchanges. It should be feasible to improve sidewalks and add stairways and ramps for the disabled. Coverings for the sidewalks, stairs, ramps, and stations are desirable. If the stairs are high and many passengers are elderly, escalators should be considered. Stairs and ramps should have adequate rails and lighting, easy grades, and landings at 1.8 m or at no more than 2.4 m of vertical change. The bus roadway might be raised to the limit permitted by the bridge clearance to reduce passengers' vertical movement. Most city buses are less than 3 m high.

A bus station can be built on a frontage road at a location without a separation, possibly to provide service to a major installation on that side of the freeway. Slip ramps for buses can be located between the freeway and the frontage road. If patrons on both sides of the freeway are to use the station, pedestrian bridges with stairs and ramps will be necessary on both sides of the freeway and frontage road stations on both sides of the freeway might be necessary.

Buses can exit the freeway to local street stations or bus stops and return to the freeway. This inexpensive method takes only a little more time than using a median or frontage road station stop if local traffic is light. Heavy local traffic, however, will have a serious impact on the bus schedules. If the buses run on the freeway lanes, stations should be at least 3.5 km apart to allow the buses to operate on schedules that match the prevailing freeway speed. With careful design, different station types that are appropriate to local conditions can be built on the same freeway.

Bus turnouts from the right freeway lane should provide substantial space for deceleration, standing at stations, and acceleration. A loaded bus has low acceleration, so these lanes must be extra long to give the bus enough room to merge into the freeway traffic. Deceleration lanes of normal length are suitable. The standing area should be at least 6 m wide to permit a moving bus to pass a stalled bus. The distance between the freeway shoulder and the bus turnout can be as small as 1.2 m in difficult cases, but a distance of 6 m or more is desirable. The passenger-loading platform should be at least 1.5 m wide, although a preferable width would be 1.8 to 3 m, depending on the number of waiting passengers. The platform length should be one bus length, more if multiple bus lines stop at the turnout.

Buses can stop in the right lane of urban surface streets, but a turnout is desirable if space is available. Turnout length depends upon the street's purpose. For arterials, a short deceleration lane should be provided. It should have a 5:1 taper plus a bus length and enough standing space for the maximum number of buses likely to use the turnout at rush hour. The acceleration lane should be long enough for the bus to attain the usual speed of the traffic. The loading area's width should be at least 3 m, preferably 3.6 m, and its length should be about 15 m per bus. Even on streets with lower traffic levels, loading buses should be kept out of the traffic stream. Deceleration and acceleration lanes can be minimal and standing space limited to one or two buses. The passenger-loading area is usually the sidewalk.

Turnouts can be at the beginning, middle, or end of the block. At the beginning or end of the block, the cross street can be all or a part of the acceleration or deceleration lane. The bus-stop locations should be coordinated with connecting bus stops to minimize the street crossing that patrons must do.

On rural arterials, shoulders usually are inadequate for bus parking and loading. Widening the shoulder to at least 4 m or providing a special turnout that provides loading areas for transit and school buses should be considered.

Bus management is the art of having enough buses on the street in rush hour to meet the demand of passengers, including standees, on routes convenient to them. It is complicated by buses breaking down in service or being unavailable during repair, the variation of rush-hour loads and occasional special heavy loads, and the necessity of forecasting loads for future bus purchases. Future loads are difficult to predict because of the effects of competition, fare changes, and general economic conditions on bus patrons.

Schedules are spaced to avoid standee overloads using the current load and size of available buses. Before and after rush hour, the goal is to provide seats while avoiding running nearly empty buses; if the time spacing is too long for reasonable service, switching to smaller buses might be advisable. Obviously, this is an art that involves mathematics, skill, and luck.

LIGHT RAIL

We used to call them horse cars, and they became streetcars after they were electrified. Some of them became light interurban trains before they were almost all retired in favor of diesel buses. Now they are making a comeback, sometimes on separate right of way and sometimes as trains instead of single cars. The comeback and the separate right of way are the results of city traffic gridlock and declining air quality. Yes, when you make a mistake you should correct it and move on.

Some recent examples of streetcar and interurban revival are the Tijuana Trolley train from San Diego and a single car line traveling eastward from downtown Portland, Oregon. Light rail differs from heavy rail in many ways. In addition to sometimes using single cars and not always being on dedicated right of way, light rail is nearly always on the surface, costs less, operates at lower speed, and makes more frequent stops.

HEAVY RAIL

The heaviest transit volumes are served by heavy rail. Most heavy rail projects are located in large, densely populated cities in developed and developing countries. A successful system requires a dense population over a large area and enough prosperous passengers to pay for the expensive installation.

In the dense city centers, the trains are put in tunnels or overhead. Tunnels can run under a surface street or freeway, or they might need to be deeper to avoid the many utilities buried in our cities. Deep tunnels require deep stations with ventilation, escalators, and elevators. Usually tunnels are paired side by side for economy, but sometimes design considerations or soil conditions require separate tunnels. Overhead construction requires a continuous bridge, usually over a surface street. Urban stations are close together for easy pedestrian access.

In slightly less populated areas with less expensive rights of way, the trains run on the surface on a dedicated right of way or in a freeway median. Passengers must have adequate access to stations from a local traffic generator, off-street or garage parking, or transfers from bus, auto, or taxi. A typical median installation is 9 m wide with 0.6-m barriers on each side. Track centerlines are 2.4 m from the barriers and 4.2 m apart.

Suburban stations are farther apart. This speeds up the schedule in relation to the competitive auto traffic and makes transit more attractive.

Off-street parking lots usually provide parking at suburban stations. The number of parking spaces depends on the other available parking facilities in the area, the number of passengers, and the number of them who walk, ride bicycles, or are dropped off by automobile or bus. Twenty to sixty places close to the station should be reserved as an area for the drop-off of people with disabilities, taxis, kiss-and-ride, and pick-up. A bus loading area should be at least 6 m wide to allow standing buses to be passed. Parking that meets the requirements of the Americans with Disabilities Act should be near the entrance, and bicyclists should have a place to park and chain their bicycles. Pedestrian access space should be clear of vehicles, and pedestrians should have to walk no more than 120 m.

Parking spaces for subcompact cars should be 2.4 m by 4.5 m. Parking spaces for full-size cars should be 2.7 m by 6 m. Sidewalks should be at least 1.5 m wide, and loading areas should be at least 3.6 m wide. Grades for roads the buses will use should not exceed 7 percent, and acceleration lanes should not exceed 4 percent. The bus roads should be at least 6 m wide, and the bus loading area should be 7.2 m wide to allow moving buses to pass standing buses. The minimum length should be 29 m for two buses and 14 m or more for each additional bus.

Because heavy-rail construction accommodates heavy traffic volumes and requires a high volume of riders to pay for it, extending lines to less densely settled suburbs requires policy decisions and possible financial subsidy. Frequently, a system is not viable financially or politically without such line extensions, and it is not viable financially with the extensions unless policy decisions are made. Some of the policy decisions can include the following:

Zoning for high density near suburban rail stations

Planning apartment complexes on public property near stations

Building large automobile parking lots that provide easy pedestrian access to the stations

Providing bicycle lockers at stations or permitting bicycles on trains

Building bikeways that lead to stations

Realigning bus routes as feeders for the trains

Providing kiss-and-ride automobile access to stations

Setting fares for outlying areas below cost to encourage transit use

Making transit fares deductible from taxable income

Providing subsidies

Prohibiting automobile entry into the city center

Raising taxes on city center parking garages and raising parking-meter charges.

COMMUTER TRAINS

Long a major factor in New York City and Chicago, commuter trains still are developing as an alternative to automobile travel from distant suburbs or exurban cities and towns. For example, Washington, D.C. started a commuter train from Harper's Ferry, West Virginia, to Union Station in the city. Union Station has a transit station as well as the railroad station.

Later, a commuter train was instituted between Baltimore and Washington, serving large and small cities between them. More recently, commuter service began between Washington and Fredericksburg, Virginia. Virginia Railway Express is the nation's second-fastest growing commuter line, after the one in Dallas. VRE grows by making customer service a priority; it pays day-care fees and issues free tickets if the train is a half-hour late, sends e-mail alerts about problems and changes, and offers free parking and a café car.

These commuter trains use the tracks and stations of the regular railway lines through a rental or operating agreement. Although they are slow-speed tracks used primarily for freight, the trains' speed is comparable to free-flowing highway speeds; this is not bad in an area that has the nation's second- or third-worst commuter traffic. The commuter trains provide additional safety and comfort in a high-traffic area.

CHAPTER 17
RAILWAY ENGINEERING

CONSTRUCTION

Standard gage is 1.4347 m, measured between the inside rail surfaces 15.9 mm below the top of the rails. The minimum subgrade width is 3.353 m on each side of the centerline, with a minimum distance of 4.267 m between the centerlines of sidings and multiple tracks. With these dimensions, the subgrade top for two tracks is 3.353 m + 4.267 m + 3.353 m = 10.973 m (see Figure 17.1).

A comparable two-lane highway with 3.35-m lanes, 1.83-m shoulders, and far less capacity has a subgrade width of 11.58 m.

Sub-ballast is 0.305 m, and top ballast 0.305 m. Ballast slopes are 1:2 and the ballast top is 2.9 m across. Ditch depth is at least 0.305 m below the subgrade shoulder; the ditch bottom is at 0.610 m. Minimum slopes are 1:1.5 for earth, 1:0.5 for broken rock, and 1:0.25 for solid rock. The ties are 2.6 m long × 0.17 m × 0.23 m.

Insulated rail plates with spikes hold the ties and rails in position. Rails, plates, ties, and ballast spread the load to the subbase. Sub-ballast is a finely graded material that keys together under compaction. Top ballast can contain material up to 40 mm.

Like the highway curves described in Chapter 1, most horizontal railway curves are simple circular curves, and vertical curves are parabolic. There are differences between highway and railway curves, however, because of the steel rail's firmness and the railway's need for more precise vehicular control.

Horizontal curves begin and end with spirals, making a gradual transition from tangent to curve. Compound curves should only be used when they are unavoidable in difficult terrain; speed should be controlled depending on the severity of angular change. Reverse curves can be used on low-speed passing and yard track, but curves on the main line must be separated by spirals and at least 30 m of tangent.

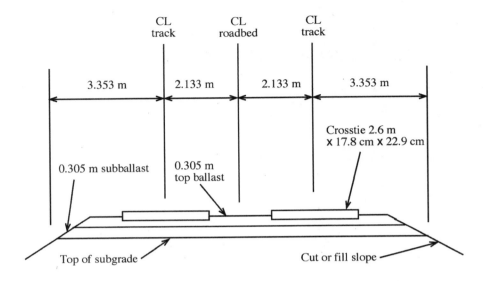

FIGURE 17.1 Typical minimum two track roadbed.

Superelevation, the relative height of outside rail over inside rail on curves, varies from 0 to 150 mm. Engineers decide the superelevation based on curvature, speed, balance, and spiral length. The 150-mm maximum is set to prevent overtilting of stopped trains, but it would cause excessive high-speed wear of the outside rail and require curve speed limits. Too much superelevation causes wear on the inside rail. The equilibrium speed is calculated using the following formula:

$$V^2 = 213.66 \times S/gD$$

where V = Equilibrium speed in km/h
 S = Superelevation in mm
 D = Degree of curve
 g = Gage of track in meters.

The speed that is usually permissible is calculated from the same formula using $S =$ superelevation in mm + 75 mm. This gives a speed that would be in equilibrium if the superelevation were 75 mm higher than it actually is, permitting moderate wear on the outside rail. Permissible speeds can be reduced for freight cars with high centers of gravity and increased for passenger cars with anti-roll devices. Equilibrium speed for a tilt-train would add the amount of tilt to the actual superelevation.

The minimum vertical-curve length is controlled by maximum grade-change rate. That maximum is 0.328 percent per 100 m for summits and 0.164 percent per 100 m on sags.

For convenience and simplification, use the next even-numbered station to calculate the vertical curve. If the grade change requires at least 6.7 100-m stations, use a vertical curve of eight stations. Add beginning and end elevations, and divide by two to find the average elevation.

The center offset from tangent is one-half the difference between the grade intersection elevation and the average elevation of the end stations. The offset at the end stations is 0 because they are on the tangents. The offset at other stations is the square of the proportional distance toward the center of the curve. At Station 1 or Station 7 on an 800-m curve, the offset is $\frac{1}{16}$ × the center offset. At Station 2 or Station 6, the offset is $\frac{1}{4}$ × the center offset. At Station 3 or Station 5, the offset is $\frac{9}{16}$ × the center offset, and at Station 4, it is 1 × the center offset. Intermediate station elevations are calculated with the same formula.

EFFICIENCY OF OPERATION

Railway transportation of passengers and freight is the most efficient mode in the following ways:

Land requirement is least

Power requirement per unit is lowest

Air pollution per unit is least

There are fewer accidents with fewer fatalities, injuries, and instances of property damage

Friction is much lower at the line of contact between steel wheel and steel rail than at the much larger footprint of tire on pavement, so moving railway traffic requires less power. The greater strength of steel rails and wheels permits heavier loads, and the virtual elimination of wandering allows for a narrower roadway and less distance between passing vehicles.

Minimizing the number of powered vehicles by using larger and heavier engines reduces the air pollution caused by rail transport. Per kilometer per ton, rail transport causes a fraction of the air pollution that other transport modes cause. Coal-powered rail revolutionized freight and passenger transport, and diesel engines reduced emissions while maintaining the same economy and speed. With larger engines, electrification or a change to alternative fuels such as natural gas or hydrogen will be more efficient in rail transport than in other modes.

Because of their professional operators, careful inspection by railway companies, and governmental oversight, trains are our safest major transportation mode. Continuing to separate traffic at rail-auto crossings enhances safety.

PASSENGER CAPACITY

Intercity passenger trains that have twenty 90-passenger cars and travel on double track at 10-minute intervals will carry 10,800 passengers per hour in each direction. This is more than five times the maximum lane capacity of automotive freeways. Station time is a lesser factor in maintaining time intervals, and greater capacity is available when cars are added or intervals between trains are reduced. This increase is easily accomplished; increasing highway lane volumes above 2,000 vehicles per hour, however, quickly leads to overload and congestive breakdown. At an occupancy rate of 1.2 passengers per vehicle, 10,800 rail passengers would fill 9,000 vehicles per hour; this is six times the efficiency of highway lanes with high-speed traffic.

The six trains carrying 10,800 passengers per hour are equal to 54 aircraft carrying 200 passengers per hour. Only five to ten minutes are required for train loading, compared with the one-hour show-up time for air travel, so the first train would be 70 miles out before the first aircraft took off. The aircraft has higher speed, but that advantage does not show up until the trip exceeds about 200 miles or 320 kilometers. Time savings can be significant when a trip is longer than this, especially for business travelers. With 10 minutes of loading time and an average speed of 110 km/h, a 320-km trip will take three hours by rail. With a one-hour show-up time, a 30-minute taxi ride at each airport, time allowed for take-off and landing, and a 30-minute flight, the same trip can also take three hours by air. In energy expended per passenger, however, there is no comparison.

Convenience and price are the competitive factors when comparing rail and auto travel because speeds are similar. For trips longer than 100 km, the quick convenience of the auto is less important, and the ease and safety of rail travel is more attractive. Rail travel for one person costs less than automobile travel; when two are in a car, the cost is almost the same, and when a car has three or more occupants, the auto is less expensive.

When transporting freight, air transport is better only when the freight is time-sensitive. Using rail freight is less costly than using trucks, but trucks have a larger market share due to the greater convenience of pickup and delivery.

HIGH-SPEED RAIL

On May 10, 1893, steam locomotive 999 of the New York Central traveled 181 km/h (112 m/h) between Buffalo and Batavia with the Empire State Express. On June 15, 1902, New York Central started the Twentieth Century Limited from New York to Chicago, which was the best and fastest passenger train in the world at the time. Pennsylvania Railroad pioneered the GG1 high-speed electric locomotive during the depression in the 1930s and 1940s. At that time, many American railroads were running regular 160-km/h (100-m/h) schedules. In the 1950s, MKT had a 240 km/h (150 m/h) diesel locomotive, but relegated it to freight service after testing. American railroads had begun to default on passenger service, and the automobile and airplane had begun to take over the abandoned market.

American railroads began with massive investments from British financiers. As the nation opened up and absorbed its vast western region, the railroads soon received a boost from a spectacular U.S.-government subsidy: Every other square mile adjacent to the tracks across the great plains and to the West Coast. The populous West Coast soon had far better contact with the nation than the clipper ships had provided, and farms, ranches, towns, and cities were founded in the interior. Railroads were profitable, innovative, and successful; too soon, however, the railroad industry matured, and owners were accused (and sometimes convicted) of price gouging, worker mistreatment, inattention to passengers, and stock manipulation.

Despite the railroad's advantages of efficiency and low cost, service declined. Passengers chose the greater convenience, speed, and comfort of auto and air travel. Trucks, partly aided by federal and state government subsidy, took away a large majority of the freight market. The subsidy provided for air transport is air-traffic control and other services that cost billions more than the user fees that are collected. The myriad subsidies provided for autos and trucks include road maintenance, research, employer tax exemption for employee parking, and parking-meter fees that are far under cost. The one-time railroad land subsidy was provided long ago, but some railroads are still major landowners.

High-speed trains are not a new technology. They only require careful improvements on current technology: Continuous welded rail, smooth roadbed, aerodynamic design, electric motors, and straight or gently curving alignment. Different nations have worked together to improve each succeeding generation and to provide engineering solutions to observed problems and limitations.

In 1964, the Japanese Shinkansen (Bullet Trains) achieved scheduled speeds of 250 km/h. The Southeast Line of the French Train a Grand Vitesse (TGV) opened in 1981. In 1989, the Atlantique Line opened with the fastest train in service. With more aerodynamic design and 54 percent more power than the Southeast Line, Atlantique has run schedules of 300 and 350km/h on three main lines and has run tests at 515 km/h. When TGV is compared with airlines, TGV uses 10 percent of the energy per passenger and is superior in comfort. Swedish, German, Italian, and Spanish trains run at comparable schedules. All of these trains run on special, high-speed track, but they can run on regular track if it is electrified. British trains routinely achieve speeds of 200 km/h on regular, electrified track. Engineers in Sweden developed an undercarriage that is capable of tilting to achieve higher speeds on regular track curves.

Train lines that stopped passenger train service because it was unprofitable need to know that full service pays dividends. The first high-speed Japanese train was profitable in its second year, and the train system (although not each line) has been profitable since. A TGV representative told Congress that the total cost of taking a TGV from Paris to Lyon, even including interest and depreciation, is less than the cost of jet fuel alone for an Airbus. The TGV is paying a 15 percent profit after debt service, and it is paying off debt early. Fast, safe, and inexpensive trains have revolutionized internal travel in Japan and France.

The French government guaranteed loans for the TGV Southeast Line, but they cost nothing because they were paid in full by fares. The government paid for 30 percent of the Atlantique Line's construction.

The Atlantique TGV is powered by 25 kv, 50 Hz/1.5 kv dual-current power, synchronous motors rated at 4,400 kw with an auxiliary power supply. Gare Montparnasse and Gare de Leon in Paris have been renovated for TGV service. TGV North has links to the Southeast and Atlantique lines, a Paris bypass, Charles de Gaulle International Airport, and Euro-Disneyland.

Spain is converting from a broad-gage to a standard system in a nationwide program that includes three major TGV lines. The remaining system is being upgraded to 200-km/h standards. The following nations are upgrading track and planning fast trains to reduce fuel-inefficient and air-polluting travel: Portugal, Greece, Turkey, Denmark, Belgium, the Netherlands, Switzerland, Austria, Ireland, Australia, Taiwan, and Korea. We are planning moderate speeds that could have been reached 50 years ago between Boston and Washington.

Napoleon first proposed an English Channel tunnel in 1802. In 1986, Margaret Thatcher and Francoise Mitterand agreed that a consortium could construct the Eurotunnel, and French and British work crews met under the channel on December 1, 1990. Supertrains now travel between Paris and London in three hours, a trip that formerly took seven hours or more. The British link is an upgraded but slower track to a new terminal at Waterloo Station.

The concessionaires' contract runs until 2042 and has a no-competition clause that expires in 2020. Financing was $8.3 billion in bank loans and $1.9 billion in equity sold to about 500,000 stockholders. A $3.8-billion overrun increased debt and payout time.

The TGV version in the Eurotunnel is a 100-km/h Transmanche Super Train, the TMST, which has 18 cars and 794-passenger capacity. There is a tunnel in each direction and a small service tunnel between them. Double-decked and enclosed auto trains carry passengers and autos from Folkstone to Calais in 33 minutes.

British High Speed Trains (HST), 200-km/h (125-m/h) diesels, or 225-km/h (140-m/h) electrics have cut travel times enough to unload congested airports. Traffic at Manchester Ringway airport dropped more than 20 percent after the HST initiated Manchester-London service. German ICE trains tested at 405 km/h (252 m/h) and run schedules at 290 km/h (180 m/h). With larger cars than the TGV and 14-car trains, ICE service started on renovated track on June 2, 1991. A system of 2700 km (1670 miles) of new high-speed track and many more kilometers of improved track is planned. Italy's high-speed rail system will link an east-west line at Turin-Milan-Venice to a north-south line at Milan-Florence-Rome-Naples and across the Messina Straight to Sicily. First service began at 250 km/h (155 m/h) between Milan and Rome.

Meanwhile, rather than see passenger train service abandoned, Congress created Amtrak. With no track of its own, Amtrak must rent space on less-than-adequate track. Top speed on this rented track is just 130 km/h (80 m/h) or less, it is maintained by others, and Amtrak is in competition with heavy freight for space. Unable to break even under such circumstances, Amtrak relies on a grudging federal-budget subsidy, which is subject to review with each new budget.

Nonetheless, Amtrak has developed the Metroliner Express between Washington, D.C., and New York, traveling at a top speed of 200 km/h (125 m/h) and a schedule of 149 km/h (87 m/h). From city center to city center, this system beats air travel for cost, speed, and especially comfort. Other Amtrak trains are comfortable and convenient, but none approach the service level provided by the nearly half-century-old Japanese Bullet Trains, much less the 10-year-old European trains.

As mentioned earlier, a common air-travel trip requires 30 minutes of travel to the airport, a one-hour show-up time, one hour of loading, time allowed for takeoff, landing, and unloading, and 30 minutes of travel from the airport to home or a business destination. A common train trip requires 30 minutes of travel to the station, 15 minutes of loading and unloading, and 15 minutes of travel to the destination. Assuming the train has a two-hour head start and the aircraft is traveling at 800 km/h, approximate competitive distances are shown in Table 17.1.

TABLE 17.1 Competitive Distances for Train/Air Travel

Train speed, km/h	200	250	300	350	400
Competitive distance, km	500	750	1000	1300	1800
Time hr	2⅔	3	3⅓	3⅔	4½

The conclusion is that TGV-type trains are competitive with aircraft to 1000 km (620 miles). When comfort and environmental factors are considered, they are competitive for many passengers at 20 percent beyond that. The Japanese experience is that the 200-km/h Shinkansen between Tokyo and Osaka, a 515-km distance, has captured 80 percent of the market. Future generations of fast trains can be even more competitive, but limiting factors might be air resistance, friction, vibration, or sound.

If airlines were relieved of their most burdensome and least profitable routes, they would be able to improve service for their long-line customers. The true limiting factor for high-speed, steel-wheeled trains might be better airline competition.

MAGLEV

No one knows how fast vehicles with Magnetic Levitation, or Maglev, will go. Speculation is that air resistance will limit speeds to about 1600 km/h (700 m/h). Other speed restraints are possible, but unlikely. Possible speeds, starting with tested speeds, and Maglev distances compared with air distances are found in Table 17.2.

TABLE 17.2 Comparitive Maglev/Air Distances

Maglev km/h	450	500	550	600	650
Distance km	2150	2700	3600	4500	7150
Miles	1340	1680	2240	2800	4450
Time hr	4.5-5	5⅔	6½	7½	11

Maglev's technology is likely to be less expensive, less polluting, and safer when compared with air travel, and speeds are comparable. Maglev at 650 km/h has the potential to compete with air travel for all continental transport except on the longest east-west trips in Eurasia, on north-south trips in America, and on over-water trips.

Magnetic levitation was envisioned by U.S. rocket pioneer Robert Goddard, who wrote in 1909, "The cars might be held in suspension by the repulsion of opposing magnets. When thus isolated, they could be propelled by the magic power of magnetism."

The first successful model was made in an American laboratory, but research fell to budget-cutters in the 1970s.

The Germans and Japanese picked up the ball and have made great strides toward a commercial version of the laboratory craft. Transrapid, the German Maglev, is almost a commercial reality. Both low-speed transit and high-speed intercity versions are in demonstration. Levitation can be accomplished either by attraction or by repulsion. The alternating push and pull of magnets with rapidly alternating polarity accomplishes propulsion.

Superconductivity, the ability of a substance to carry electricity without resistance, has the promise to enhance the already proven methods of levitation and propulsion and to revolutionize electrical transmission and many other industries. The first superconductors required cooling near absolute zero in liquid hydrogen. More recent discoveries have used liquid nitrogen as a coolant, and research continues.

CHAPTER 18
PEDESTRIAN FACILITIES

SIDEWALKS

Pedestrians are the lifeblood of commercial enterprises in urban areas, but it is difficult to resist the overwhelming demands of automotive traffic. This is true even though drivers are typically pedestrians more often than they are motorists. We change roles so rapidly! Pedestrian-friendly design must be done carefully to give flesh and blood a chance to compete with tons of steel. Inadequate sidewalks force pedestrians into the street with motorists.

Peak flow on urban sidewalks occurs at noon, whereas peak traffic flow on the streets usually occurs at mid-morning and mid-afternoon. The maximum flow is about 70 persons per minute per meter of sidewalk width. As with automobile traffic discussed in Chapter 6, capacity plummets when density becomes too great. (Figure 18.1).

Maximum flow occurs at a walking speed of about 40 m per minute. Computations of required sidewalk width should include an additional 500 mm for the proximity of a wall and a full 1 m if the wall includes display windows. Include also the width of the sidewalk area occupied by street hardware such as parking meters, fireplugs, mailboxes, and newspaper boxes.

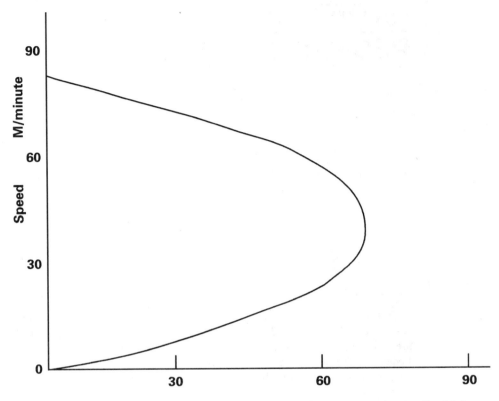

FIGURE 18.1 Pedestrian speed and flow.

A 6-m sidewalk that is located next to a display window and has 0.5 m occupied by street hardware would have the capacity for the following number of pedestrians:

4.5 m (70 persons per meter per minute = 315 pedestrians per minute.

Peak flow occurs near the boundary between C and D levels of service. The following are the levels of service:

Level of Service A: 12 m^2 or more per person. Pedestrians are free to choose a walking speed and have little conflict with others.

Level of Service B: 4 to 12 m^2 per person. Pedestrians are aware of others and give them some accommodation.

Level of Service C: 2 to 4 m^2 per person. Pedestrians must make minor speed and direction adjustments to avoid conflict with others.

Level of Service D: 1.5 to 2 m^2 per person. Pedestrians have restricted freedom to choose speed and direction, and frequent changes are required.

Level of Service E: 0.5 to 1.5 m² per person. Speed is reduced, and cross-traffic movement is difficult.

Level of Service F: Less than 0.5 m² per person. Pedestrians usually are stopped and able to move only when others move. Frequent and unavoidable contacts occur.

Sidewalks along rural roads reduce accidents and save lives. This is especially true at community facilities such as schools, churches, stores, and recreational facilities, and along roads between homes and those facilities. Lack of a sidewalk is hazardous even when pedestrian traffic is light. We need to be keenly aware of this even when economics and greater competing needs force us to decide against sidewalk construction. In some cases, a wider shoulder is a valid compromise. In others, pedestrians might use a trail that doesn't necessarily adjoin the road. Usually, the close calls are left to the engineer's judgment, and we should err on the side of the pedestrian, who is the most vulnerable.

Suburban sidewalks, which were once always in the plans, now sometimes are deleted. This circumstance forces people to make a choice between driving and walking in the street with traffic. Those who are too young or too old to drive will not have the choice. Is this what we want for our most vulnerable kin? This trend should be stopped in its tracks. Sidewalk width can vary from 1.2 m to 2.4 m, depending upon the expected traffic volume. An additional 0.6 m should be provided as separation from auto traffic if the sidewalk is adjacent to the curb. Sidewalks always should have all-weather surfacing; otherwise, people will walk in the road.

Sidewalks wide enough for pedestrian traffic should be placed on both sides of the street if they provide access to schools, parks, shopping areas, transit stops, or commercial areas. The sidewalks should be at least 2.4 m wide, wider if necessary. In residential areas, sidewalks need to be on at least one side of the street, but it is better to have them on both sides. The sidewalks should be placed well away from traffic, preferably near the right-of-way line.

STREET CROSSINGS

Pedestrian crossings at busy urban streets should be kept to a minimum, but they cannot be prohibited. Providing better choices can reduce pedestrian crossings in these dangerous conditions. Wide urban arterials are especially dangerous for pedestrians, especially the elderly or disabled, who often move at a slower pace. Pedestrian protection should include fixed-source lighting, refuge islands, barriers, and signals. The walk signal and refuge island are especially good at promoting safety when pedestrians must cross multilane arterials that have frontage roads with long distances across. A refuge island is an area near the center of a wide street. It is protected by barrier curbs and is wide enough to reassure pedestrians caught in mid-crossing at a light change.

Downtown Minneapolis and Duluth have numerous second-story pedestrian bridges to facilitate shopping, promote safety, and protect pedestrians from the weather. They are worth emulating when streets are busy or narrow, even by those who don't have the compelling reason of frigid Minnesota winters.

At intersections, the sidewalk must provide sufficient space for the pedestrians waiting for the walk signal, plus enough space for other pedestrians to pass them. The crosswalk must be wide enough to accommodate the waiting pedestrians in both directions within the time allotted for the pedestrian phase. This means the crosswalk is not limited to the width of the sidewalk; in heavy pedestrian traffic areas, it should be wider. Impatient motorists wishing to turn across the pedestrian stream make it difficult for pedestrians to clear the street within the allotted time. Moving back the traffic stop line instead of moving closer to the parallel street is the best way to widen the crosswalk.

Making sidewalks safe for people with varying disabilities requires diverse planning actions. The visually impaired might need larger street signs; those with greater vision impairment might need texture near the curb, especially where curb cuts make it difficult to locate. Pedestrians in wheelchairs can be fully mobile if adequate curb cuts are built, but they must depend on others where the curb is insurmountable.

Curb cuts should be located well within the crosswalk, not near the auto traffic on either side. Curb cuts should also be made when pedestrian crossings are installed at midblock, unless a nearby alley entrance performs the same function. Visibility should be unimpaired and warning signs should be posted. Curb cuts and safety islands should be at least 1.2 m wide. A wheelchair pushed by an attendant requires a 1.8 m width, and that should be considered where it is feasible.

PEDESTRIAN BRIDGES

Separations for pedestrians are justified when volumes of pedestrians and vehicles in intersecting paths are great enough to interfere with ready passage. Such situations can create major safety hazards, substantial congestion, and time loss. Separations are warranted where heavy peak pedestrian traffic from generators such as schools, athletic fields, factories, office buildings, commercial areas, or central business districts must cross streets with moderate-to-heavy traffic. Lighter pedestrian traffic crossing streets with more severe auto traffic can also qualify, but first there should be a search for a less drastic solution. An athletic field might be placed on the right side of the highway, for instance, or the pedestrian walkway on a nearby street separation might be improved.

As noted in Chapter 15, Transit, pedestrian bridges often are needed in conjunction with bus passenger stations in freeway medians or on busways parallel to the freeway. Also, street overcrossings of freeways near bus stations are expanded with pedestrian facilities to provide access to the ground-level station.

Frontage roads parallel to a highway can be included in the length of the pedestrian overcrossing if the frontage road traffic justifies the addition. If the traffic is light and there is sufficient room for stairs and ramps between the highway and frontage road, the pedestrian overcrossing of the highway should be sufficient.

Some undercrossings have become crime scenes, and it is important to provide good vision through the tunnel. Gently slope the walkway down to the entry and exit and provide adequate light, especially in long tunnels. Fencing is necessary on both overpasses and underpasses to assure the safety of pedestrians. The unsafe decision to base actions on convenience instead of safety should be strongly inhibited.

Overpasses should be equipped with screening to keep objects from dropping onto the traffic, although this cannot be completely prohibited. Small objects can be pushed through netting or thrown over the top of the fence. If the fence is covered to put the walkway in a tunnel of netting, children may climb on top of it. We have found no complete answer other than "try your best."

The following situations can indicate the need for protective screening:

When responsible adult or police surveillance seldom will be present

When unaccompanied children will use the overcrossing frequently

When vandalism has occurred in nearby areas.

Overcrossing and undercrossing walkways should be at least 2.4 m. This minimum width will accommodate a substantial number of pedestrians. Wider walkways should be considered in areas where pedestrian volumes are high, such as in a city center or near a large school or employment center.

The design of pedestrian overcrossings and undercrossings is in the province of Structural Engineering, which is another book in this series.

CHAPTER 19
RIGHT-OF-WAY

RIGHT-OF-WAY PLANNING

After planning of the construction project, and the field layout of the roadway cuts and fills, a right-of-way plan can be drawn to show the property to be taken. The right-of-way line must encompass land that is needed for earthwork, future construction, noise barriers, landscaping (including trees and shrubs to screen the road from residences), drainage construction, and fencing.

Fencing along the right-of-way line can keep unwanted vehicles, pedestrians, or animals from accessing the road. Local conditions must be considered when establishing the right-of-way plan. For example, vegetation might need to be cleared inside curves to allow for sight distance, or more room might be needed for snow storage or flood control.

Flood control is important because pavement construction accelerates runoff. The containment dam described in Chapter 3 holds siltation on site and retains initial water flow. This dam meets the agency's responsibility to preserve water quality and to prevent flood damage. A wide right of way permits the construction of such containment dams, as well as gentle slopes that make the road safer and more comfortable for drivers.

The right-of-way plan should show parcel ownerships and numbers for property that is within or partly within the right-of-way line.

SCHEDULING

The schedule should include the appraisal, negotiation, and acquisition of any private property that is required for the construction. Many project schedules have gone awry at this point when management failed to recognize the complexity and difficulty of the acquisition process.

The appraisal of the parcels is a straightforward process with standard procedures, but it has plenty of room for differing opinions. The process involves describing the property, evaluating its condition, finding recent sales prices of usually three comparable properties, evaluating the differences among these properties, and stating an opinion about the property's value and the value of what will remain after the taking. Frequently, two independent appraisals are made. A third appraisal is added if the first two differ substantially.

After the appraisals, an agency negotiator approaches owners of property scheduled to be taken in whole or in part. Owners react differently to the news their property is targeted to be taken. One owner might be glad to sell, while another might be resigned to a role of helping meet the community's needs and hope to get the best price possible. Owners also could be unhappy because their own plans are being disrupted or angry at the prospect of losing property they want to retain. Negotiation procedures differ somewhat from state to state because state laws differ, but there is a basic similarity. The government's right of eminent domain and the property owner's right to due process of law are inherent in the negotiation process. The owner can challenge the necessity of the taking or challenge the necessity of the project itself.

The negotiator usually has little leeway in negotiating the price but the project's details are more negotiable. An agreement might be reached to construct a better access road to the remainder, to take less of a certain area, or to make other project changes that do not adversely affect the project's purpose. Often, even with a willing seller, there is disagreement about the property's value or the remainder's value. If the parties cannot agree, the matter must be settled in court.

An attorney usually has more leeway in the settlement of a case than does the agency negotiator. Because of the cost of a court case, for example, agency negotiators' settlements are usually with owners who prefer immediate settlement, while attorneys' settlements are later but the cost is higher. This is a well-known fact that affects the attitudes of knowledgeable owners and delays project schedules.

Most settlements are likely to occur "on the courthouse steps." Court cases are relatively rare, but they have a big effect on the project schedule when they happen. Right-of-way acquisition, public consultations and hearings, and environmental planning are major parts of modern construction-project scheduling.

TEMPORARY CONSTRUCTION EASEMENTS

Sometimes property is needed for a project during the construction, but it is unnecessary for the finished road's operation and maintenance. Examples include a place for a materials stockpile, a borrow site for fill material, or a space that is cleared and leveled inside a curve for sight distance. Depending on the circumstances, the parties might agree to a rental or materials purchase; in other circumstances, a temporary easement for construction purposes might better serve the parties' interests.

An easement is a right to use property for specific purposes. An easement for a buried pipeline, for example, gives a permanent right of occupancy and periodic entry for service and maintenance. A temporary easement sets an ending date or event for the easement.

With a temporary easement, title does not pass, and the property remains on the tax rolls. In fact, the title is not encumbered, as it would be with a permanent easement. By agreement, a property might be improved during a temporary construction easement; for example, a hill in a field might be leveled and the stockpiled topsoil spread and smoothed.

PERMANENT ROAD EASEMENTS

Sometimes an owner prefers to retain ownership of the property, but will allow construction and use of the road. A wheat farmer, for example, might want to farm to the ditch line to control weeds and have the extra bit of crop land. A Native American tribe or another major landowner might have no objection to the road but would want the land back when the road is no longer wanted.

For the road agency, losing control of a small, unimportant road element (fencing, for example) is balanced by less maintenance and probably less cost than outright purchase, and the property remains on the county tax roll. For the road purposes that are the agency's aim, the easement gives all the rights that ownership would, with one difference: utilities would have to get their own easements.

JOINT USE

Except in areas where property costs are very high, the right-of-way plans for new roadways and, where feasible, the plans for rehabilitating old roadways should include provisions for future construction in that location. This includes future lanes, medians, shoulders, cuts and fills, ditches, landscaping, sidewalks, utility strips in border areas, and outer slopes. The space made available for utilities should be well away from normal road operation and maintenance. It should provide sufficient space for all utilities that might use the right of way.

The road or street often must serve as a utility corridor, and it sometimes is required by ordinance to do so. Road are desirable locations for utilities because roads go where the people are, as the utilities do, and provide convenient access to service, maintenance, and replacement when technology changes. Utilities prefer not to deal with each property owner along the route, and the property owners certainly do not want to deal with a dozen utilities.

The utilities that might be installed along a road or street include:

Sanitary sewers

Water mains

Oil, gas, and petroleum-product pipelines

Overhead and underground power and communication lines

Drainage and irrigation water lines

Heating mains

Tunnels that connect buildings.

Agreements for joint use of a road's right of way should specify the rights and responsibilities of both parties. The agreement's language should be specific enough to prevent most controversies but general enough to avoid frequent renegotiation. Utility construction, maintenance, and replacement in the roadway disrupt traffic. Utilities should be installed at least 0.5 m behind the curb, preferably behind the sidewalk or near the right-of-way line.

Utilities should not be constructed under the road except at crossings. In special circumstances, such as urban-center streets that are the only feasible utility locations, strict agreements must detail responsibility for traffic disruption and street repair made necessary by utility work. Special consideration and treatment should assure minimal traffic disruption during the construction, and the construction should emphasize a design that can be serviced and maintained without street damage.

At stream crossings, utilities can be attached to bridges when no other alternative is available. An effort should be made to preserve the aesthetic appearance of the bridge.

CHAPTER 20
TRAFFIC MANAGEMENT

AUXILIARY LANES

Most climbing lanes are on two-lane highways. Climbing lanes are not usually needed on multi-lane highways, because passing is not obstructed by opposing traffic. If traffic volumes are so great that queues form behind slow trucks in the right lane and vehicles have few opportunities to merge left to pass, a climbing lane might be considered, or the need for additional through lanes might be evaluated. Sometimes a climbing lane on a multilane road will be needed at the design year but might be constructed at a later date.

Truckers almost always comply with the keep-right signs for climbing lanes. Climbing lanes are required when a combination of heavy trucks, a sustained heavy grade, and heavy auto traffic greatly reduces faster-moving vehicles' opportunities to pass slower-moving vehicles. The need for climbing lanes is shown dramatically by the following statistics: Numbers of truck accidents rise from below 100 accidents per 100 million vehicle kilometers at 0 speed reduction below average running speed to the following:

- 350 accidents at 10 km/h speed reduction
- 900 accidents at 20 km/h speed reduction
- 2400 accidents at 30 km/h speed reduction.

On long, heavy grades, loaded trucks reach a critical point (Table 20.1) at which they lose speed until they retain a low-gear speed up the grade from the critical point. Approximate critical points are cited in Table 20.1.

TABLE 20.1 Critical Points for Loaded Trucks on Long Grades

500 m	for 9%	19 km/h final speed
600 m	for 8%	21 km/h final speed
800 m	for 7%	23 km/h final speed
1000 m	for 6%	26 km/h final speed
1200 m	for 5%	31 km/h final speed
1600 m	for 4%	35 km/h final speed
1800 m	for 3%	45 km/h final speed
2600 m	for 2%	59 km/h final speed

The last two critical points are not well-defined, because an increasing percentage of trucks maintain speed at grades below 3.5 percent.

When choosing a site for a passing lane, be sure to do the following:

Have a minimum sight distance of 300 m approaching the entrance taper

Avoid intersections and driveways, bridges and culverts, and areas with restricted shoulder width

Design the lane with a minimum length of 300 m when there is an isolated problem. Ordinarily, the lane should have either a minimum width of 500 m plus tapers or an optimal length of 1 or 2 km and a width that is the same as the highway lanes.

The length of the drop taper is $L = 0.6\ WS$, in which L is the taper length in meters, W is the width in meters, and S is the speed in km/h. The length of the lane-addition taper can vary from one-half to one-third times the drop taper.

Passing lanes are four-lane sections built on two-lane roads to satisfy the need for safe passing zones, to provide vehicles with opportunities to pass slow-moving trucks, or to do both. The transitions between two lanes and four lanes must be carefully signed and marked, especially if the four-lane section is greater than 3 m. This reminds drivers that the road is basically a two-lane road and care is required when passing. Transitions should be in full sight of the driver. Widening can be done with a minimal shoulder between 1.2 m and 1.8 m in width. The added lanes should be at least 3 m wide, preferably 3.3 m or 3.6 m, and the traveled way must be at least 12 m. Four-lane sections can have a median, which might be appropriate for roads that ultimately will have full four-lane construction or for roads with 500 vph or more.

A passing-lane sign should be installed 1 km from the entry. Signs at 3 km and 10 km are desirable; they inform impatient drivers that a passing opportunity is ahead and deter unsafe passing. A sign at the entry should inform slow drivers to keep right.

A turnout might be installed where there is not enough space for a passing lane or economics do not permit one. Minimum standards for a turnout include the following:

The minimum sight distance is 300 m in each direction

The available width is 5 m

The surface is firm and smooth.

The minimum length is 60 m for 40 km/h
75 m for 50 km/h
90 m for 60 km/h
100 m for 70 km/h
120 m for 80 km/h
150 m for 90 km/h
170 m for 100 km/h.

The maximum length is 200 m so drivers do not mistake it for a passing lane. The recommended minimum width is 3.6 m, and desirable width is 5 m. Signs should be posted to encourage the turnout's use.

Passing on shoulders usually is forbidden. Marking shoulders as passing lanes should be done only when a substantial need exists and when a full passing lane is not practical. Other requirements are a shoulder at least 3 m wide and 300 m long, adequate structural strength, and a paved surface.

Emergency escape ramps are needed when the road has curvature and its downgrades are severe and extended. Truck drivers can lose control of their vehicles when overheating or mechanical failure causes brake loss or when the driver fails to make an appropriate downshift. No specific guidelines exist for escape ramps, but many have been built. The goal is to provide acceptable deceleration rates and to restore driver control. The design should be for the worst-case scenario: an out-of-gear truck with failed brakes.

The resistance forces are inertial, air, rolling, and gradient. The inertial force of the truck is countered by the others. Air resistance is insignificant below 30 km/h, significant above 80 km/h, and moderate between. Air resistance is commonly neglected and assumed to be part of the safety factor. The rolling resistance (Table 20.2) is that of the roadway surfacing material in the escape ramp, expressed as an equivalent upgrade.

TABLE 20.2 Rolling Resistance of Roadway Surfacing Material

Material	Resistance, kg/1000kg GVM	Equivalent Grade
Portland cement concrete	10	1.0
Asphalt cement concrete	12	1.2
Compacted gravel	15	1.5
Loose sandy earth	37	3.7
Loose crushed aggregate	50	5
Loose gravel	100	10
Sand	150	15
Pea gravel	250	25

The ramp length required to stop a runaway truck depends on the ramp's grade and the equivalent grade of the material on the ramp. The table above shows the superiority of pea gravel. The relative acceptability of other materials depends upon the material's price and availability compared to the cost of extending the ramp and the availability of space. The table also shows the relative utility of the loose sandy earth, which is the least expensive material, and the effect on gravel that becomes compacted and has not yet been loosened.

To prevent accidents as the truck exits the highway, the escape ramp must be on the right and on a tangent or flat curve. The design should be for at least 130 km/h, preferably 140 km/h. It should be of sufficient length and wide enough for two or more trucks. Experience has shown that a second truck might need the ramp while the first truck is still on it. That width should be at least 8 m, preferably between 9 m and 12 m. Ramps have been built ranging from 3.6 m to 12 m in width.

Material used for the ramp should be clean and loose. A geotextile earth cover might be necessary to prevent earth fines from intruding into the material. The material preferably should be rounded, uncrushed, near single-sized, and free from fines. The largest size should be 40 mm. The minimum depth should be 0.6 m; up to 1 m is recommended. The depth should be 75 mm at the entry, with full depth reached in 30 m to 60 m, and the material should provide good drainage to prevent freezing. The ramp's full length should be visible before the truck reaches the entry, the angle of entry should be 5 degrees or less, and an auxiliary lane on the shoulder should be considered to facilitate entry.

An advance information sign should be installed, as well as a sign at the entry. Delineators should outline the entry and the ramp, and illumination should be installed if this is feasible.

SIGNING AND MARKING

Signs should be an integral part of a road's design. Early attention to signage eliminates many operational problems. The extent of signage depends on traffic volumes, the type of facility, and the degree of traffic control required for the roadway's safety and efficiency. Each traffic-control device fills a specific need to regulate traffic or warn, guide, or inform drivers. An engineering study, as well as the geometric design of the road or street, determines the need for signage. Warning signs have messages or symbols that point out hazards ahead. Guidance signs can provide the name of a route, the road's direction, or the distance to a destination. Information signs point out available roadside services, rest areas, or historical sites. Regulation signs give the rules of the road, such as speed limits, weight and height limits, passing, parking, and turning restrictions, and which lanes slow traffic should use. They also convey temporary requirements for traffic to merge with another lane or to wait for an escort vehicle in a construction zone.

Markings supplement, clarify, and delineate the information given on warning and regulation signs, and they point out other hazards and regulations. Three main types of markings are *pavement markings*, *object markings*, and *delineators*.

Pavement markings are centerline stripes, lane stripes, edge stripes, no-passing stripes, stop lines, crosswalk lines, safety-island markings, and markings on obstructions such as curbing that provide better visibility.

Object markings are placed on obstructions in safety zones that cannot be moved. They are painted with diagonal stripes or highly visible material and sometimes illuminated or reflectorized. Marked objects include bridge abutments, piers, poles, guardrail ends, or retaining walls.

Delineators are post-mounted, object-mounted, or pavement-mounted reflectors used where pavement markings might be inadequate in the darkness. They are found on centerlines, pavement edges, turn lanes, exits, entrances, culvert ends, safety islands, and curbs.

Products recently developed for pavement marking have longer useful lives than those that previously were available. They are much more resistant to weathering, traffic, and even snow removal.

Overhead lane signs are helpful, especially when snowfall obscures pavement markings. The sign is usually visible well before the pavement marking and permits a driver to make advance movements into the desired lane.

SIGNAL LIGHTS

The designer should integrate the initial signal system into the overall design and should define the initial system and the ultimate system. When signals are required or anticipated in the design process, their integration into the design should eliminate many operational problems, decrease maintenance, and postpone their obsolescence and replacement. The extent of signal-system planning in the original design depends upon the traffic volumes, type of facility, and the degree of traffic control required for the road's safety and efficiency.

A phase for concurrent left turns should be included when these movements will be substantial, because this is the safest and most efficient use of signal-phase time. Large numbers of automobiles are difficult to serve at a location other than the point where drivers desire to turn. Smaller movements can be diverted to a nearby left turn, or drivers can be required to make a series of right turns if through traffic makes great demands on the signal phases.

A signal system that responds to traffic needs is important, especially as traffic volumes mount. Traffic needs fluctuate by the day and by the hour, and signals that recognize and respond to current needs are the best way to save drivers time in densely populated areas. The system should include devices for detection, data processing, and control. Phase time should be allotted in accordance with demand.

LIGHT COORDINATION

Signals should be coordinated to produce practically continuous movement in the traffic's predominant direction while providing adequate service to the other movements. Signals are set for progression of three principal movements in the peak period, and another setting is used to fit the off-peak traffic flow. Free movement of the major traffic movement is an effective way to improve traffic flow on the arterial, the cross street, and the turning movements between them. When large volumes of major traffic move smoothly through the intersection, the minor cross movement and the turning movements will have openings.

Skipping phases should be provided when no traffic demand appears in a phase of the cycle. Unused phases and idle or lost green time must be held to a minimum so that the demand of the predominant coordinated flow does not unduly limit other phases.

A coordinated signal system is usually the most economical way to improve traffic service in densely developed areas. Although it is expensive, this signal control equipment is far less costly than the additional construction and right of way that would be required to achieve comparable results without it. Installation and maintenance of a sophisticated coordinated signal system, to keep it performing at peak efficiency, must be accepted as an operating cost that is just as important as the construction cost.

TRAFFIC CALMING

Traffic calming means slowing the tempo of automobile traffic to make it compatible with pedestrians, rather than requiring humans to get out of the way. It has been practiced in two environments: residential neighborhoods and city centers. In residential neighborhoods, the method of traffic calming has been to construct streets that can be negotiated only at low speed. In city centers, the method has been to restrict automobiles to the extent that human power, small vehicles, and transit have priority.

Traffic calming has been practiced in most countries, but it has been most emphasized in the Netherlands and in Germany. For more than 30 years, the Dutch have used the *woonerf*, or living yard. Cars are required to move slowly around trees and other landscaping in residential roads designed to prevent vehicle speeds much above pedestrian velocities. Non-motorized traffic has priority, and the car is welcome as a guest. The plan became popular and widespread.

Soon after the Dutch started this plan, the Germans began their *Verkehrsberuhigung* (traffic calming) program in residential neighborhoods. It has spread to thousands of residential areas and even to complete cities throughout the country.

In another program, routes scattered throughout Germany were designated as Tempo 30 (30-km/h speed limit), and nearby routes were designated as Tempo 50 (50 km/h or 31 m/h speed limit). At the end of the five-year test in 1989, the comparative records of Tempo 30 and Tempo 50 accidents, noise, and exhaust were so impressive that Tempo 30 has widespread demand.

Bremen began a traffic cell program that has been adopted by Besancon, France; Groningen, the Netherlands; Goteborg, Sweden; other European cities, and Tunis. The cell program consists of a ring road around the city center with cells established inside the ring road. Automobiles are permitted within a cell but cannot cross the cell boundaries; only public transit, emergency vehicles, mopeds, bicycles, and pedestrians can cross the boundaries. The plan has produced better public transit and fewer accidents. Many cities have plans with different methods, but they all have the same aim—to calm the auto traffic and reduce it to a scale that returns cities to the people.

PARKING CONTROL

During daylight hours, a typical American city center is thronged with automobiles, many of them searching for a vacant parking space. A motorist passing 200 occupied parking meters in the search has found an occupancy rate greater than 99.5 percent, even if the next space is available. This kind of saturation is very common.

As the throngs of automobiles become denser and spaces become more sparse, parking garages fill. Metered parking is cheaper because city governments are reluctant to raise rates. Downtown merchants need customers, and they, of course, have influence. Many of these same merchants and their clerks, who have the advantage of early arrival, fill parking spaces and feed the parking meters all day, while prospective customers search for parking. This is a case of the individual merchants' interests conflicting with the interests of the group of merchants, and thus of each individual: a classic reason laws are passed for the general welfare.

City enforcement of parking time limits is a weak link in the chain. Police resist parking enforcement; they are trained and armed for duty that is more important. Parking attendants readily identify the red meters that indicate unpaid overtime parking. Paid parking beyond the limit is more difficult to identify, and it is a lower priority. The only reasonable solution to the parking problem is to do what any merchant would do if faced with overwhelming demand for a product: increase the price.

This urban congestion solution is simple and easy—except politically. Just raise parking-meter rates and reduce legal parking times to inhibit all-day meter feeding. With a lighter workload, parking attendants can do a better job enforcing maximum parking times. Parking-garage owners, relieved of some of the unfair competition from the public sector, will raise their rates and pay more taxes. Shorter meter times will discourage meter feeding and make more space available for shoppers, and the merchants will profit.

Everyone wins in this situation, except the automobile owners who lose a part of the hidden subsidy. Even automobile owners might win if they are persuaded to use city transit as a more economical choice, or they might consider themselves winners for saving time at a reasonable cost. Traffic flows better, and city center pedestrians are safer.

One solution that is more thorough—yet more difficult politically—is to define the city center as a pedestrian zone and to emphasize parking facilities at the zonal periphery. This is the solution most of us prefer when we are pedestrians, but we don't like it when we are drivers.

Because cities differ, each must solve urban congestion in its own way. Some cities might choose to ban autos in city centers, to consider major parking facilities at the margins of the central zone, and to enhance public transit as well as bicycle storage and rentals. Some might choose to leave a network of automobile arterials within the central zone oriented toward parking garages, with or without street parking. Others might choose a zone that is smaller than the whole city center. This could be a realistic solution, or it could be a way to dodge a tough decision.

Parking problems illustrate how choices we made years ago have started to cause problems as circumstances change. They also illustrate the political nature of decisions that must accompany any effective solution. Engineers no longer can do their work as nonpolitical technicians; they must become concerned and expert citizens.

MAXIMIZING ARTERIAL CAPACITY

Urban arterials can be from two lanes for low traffic volumes to eight or more lanes for high volumes. For brevity and clarity, we will discuss a six-lane arterial, but most of these comments would apply to all arterials.

If free flow is maintained, maximum lane capacity and speed at that capacity should differ little from the freeway maximum, 2000 vehicles per lane per hour. For three directional lanes, this would be a directional capacity of 6000. Free flow, however, is difficult to attain in urban conditions. A six-lane urban arterial can have two parking lanes and frequent signal lights with left turns permitted.

Signal lights, even those favoring the arterial, seldom have more than 50 percent green, with red and amber comprising the other 50 percent. Frequently, the left-turn arrow is separate and the green for the through lane is of even shorter duration. Uncoordinated signals will reduce free flow even further, so choosing to coordinate lights is the first way to control flow restrictions.

Right turns seldom cause major interference with other traffic movements, because they are a merging movement. Free right turns, which defer to pedestrians and through traffic, help signal control. Most conflicts that occur in right turns are caused because the curb radius is too short and forces the turning vehicle to encroach on the next lane. This is a design for minor traffic that can become inadequate with traffic growth, but it usually is easily corrected. If the right-turn movement becomes substantial, a taper or short right-turning lane can reduce the conflict.

Left turns cause a major conflict between through and crossing traffic; they contribute to accidents and diminish the capacity of the road. Channelization can provide for single or double left turns. A left-turn lane alleviates the conflict with following and passing traffic, unless there is a second left-turn/ahead lane. The conflict with the opposing traffic that must be crossed still remains. Only driver awareness and signal control can reduce this conflict.

A center lane that lacks a median for left-turn channelization and storage might be encumbered for two or more cycles if the left-turn movement is a major one. Signals and left turns can reduce flows more than 50 percent in the left lane, causing, for example, a 60-percent reduction. Parking on two lanes not only reduces flow on those lanes to zero, it also reduces flow on the middle lanes as much as 50 percent because of interference from parking, unparking, and illegal double parking.

Lane switching and frustration are common when drivers must choose between delay behind left-turning vehicles and delay behind parking vehicles. When both occur at the same spot, traffic comes to a halt. This describes actual observed conditions on several six-lane arterials in Washington, D.C. With the estimated delay pattern, flow is 40 percent on one lane and 50 percent on another, reduced 50 percent by signals to 45 percent of one free-flowing lane. That is only 900 directional vehicles per hour for three lanes.

Prohibiting parking during rush hour usually clears the middle lane, but ticketing and hauling cars that are left in the curb lane is always a problem, even with cooperative police. Police, as well as local citizens, usually consider parking enforcement a low priority compared to their other duties. When rush-hour parking is prohibited, flow is estimated at 40 percent on the left lane, 100 percent on the center lane, and 60 percent on the curb lane, reduced to 50 percent by signal lights. The three-lane total is 100 percent of a free-flowing lane or 2000 vehicles per hour. Prohibiting parking at all times would free the curb lane except when there is minor right-turning interference, and it would add 400 vehicles per hour to the capacity. Prohibiting left turns with unmeasured effect on the other local streets would add another 500 vehicles per hour to the arterial capacity.

A decision might be made to concentrate traffic on an arterial to relieve pressure on others. This would be accomplished by eliminating all possible barriers within the present right of way. Taking all of the above actions, in addition to eliminating signals with separation structures and center-curb, and permitting only right turns on and off the arterial, would establish near free-flowing conditions and a directional capacity close to 6000 vehicles per hour. By eliminating the curb and reversing the center lanes during rush hour to a 2-4 configuration, the rush-hour directional capacity would be about 7500 vehicles per hour. This is more than eight times the capacity of the arterial with signals and parking.

Partial solutions might be adequate in some cases. When they are not, traffic will seek other routes and sometimes gridlock the whole system. Achieving major traffic capacity requires vigorous solutions. All solutions, including partial ones, have adverse consequences for some road users, adjacent property owners, or taxpayers. These consequences must be addressed and, when feasible, ameliorated.

A ban on parking at all times is more enforceable and is less likely to be violated inadvertently than a partial parking ban. In many cases, the lane will be needed by traffic over a span of 18 hours or more. There is no requirement to provide local on-street parking, but it might be advisable to explore possibilities for nearby on-street parking or for private or public parking facilities on vacant property.

Left-turn prohibition will free the left lane, but it will require the left-turning traffic to turn right and circle a block or more to find a crossing point. Impact on the local street system should be evaluated; it might be minimal or major. If the detour is a major traffic movement, there might be a need for signage or local street improvement. Commuters' regular trips might end in a few extra blocks of travel, but local people would quickly learn the route.

Closing some crossing streets except for right-turn entry and exit and building separations at others will eliminate signalized intersections and facilitate cross traffic, although some traffic will be required to find new routes. It will concentrate the cross traffic on routes convenient for improvement and calm traffic on some local streets. If it is necessary and feasible, some intersections can become single-point diamond interchanges. The advantages for local traffic probably outweigh the disadvantages. In a commercial area, businesses would be affected when customers have to park behind the storefront and walk around. Residents of properties fronting the arterial would experience extra noise and would be required to park behind their property. Noise barriers might be one solution, but aesthetic considerations would be necessary.

Bridge and interchange construction is costly and must be done in accordance with priorities and budgets. This clear cost will be balanced by commuter time and gasoline saved in the benefit/cost ratio. It also will be balanced by frequently ignored benefits that are not so easily estimated, such as less air pollution and less costly respiratory disease. These problems can be ignored when they are externalized: that is, when they are passed to others who might not be aware they are paying another's expenses. Our traditional benefit/cost ratio might need revision as people learn to resist externalization.

Where possible, right turns onto and off the arterial should be from an added lane before and after the intersection. This would reduce acceleration and deceleration conflicts, and it would provide a place for bus stops off the through lanes. Closing some streets to right turns could contribute to traffic calming on some residential streets if other streets are capable of handling the load.

If traffic projections warrant, a city can create a super arterial by making it a one-way street and reversing it morning and evening. Another method would be to construct a one-way pair of arterials. Either method could provide directional capacity of 12,000 vehicles per hour. A reversible arterial would work well only if the minor direction of movement was served by nearby parallel streets. As the directional split nears 50/50, reversible facilities become inappropriate.

With commuter traffic better handled by and attracted to a few improved, free-flowing arterials, local and residential streets will be relieved of much through traffic. Quiet residential streets and reduced traffic on local collector streets will make the city more livable and provide more opportunities for non-motorized traffic.

We can make our transportation system people-friendly:

By providing free-flowing traffic for commuters and others on selected streets rather than gridlock on all

By reducing traffic conflicts with street closures and separations for traffic crossing our arterials

By reducing auto volumes with better services to bicyclists on safe, dedicated streets

By improving transit with benefits commensurate with transit contributions to safety, health, and economy

By making pedestrians safe in city centers that have no room for hordes of automobiles

By calming the automobile traffic in our neighborhoods and non-arterial streets

By being open to and supportive of new methods and technologies that might solve transportation problems.

APPENDIX A
BRIDGE ENGINEERING

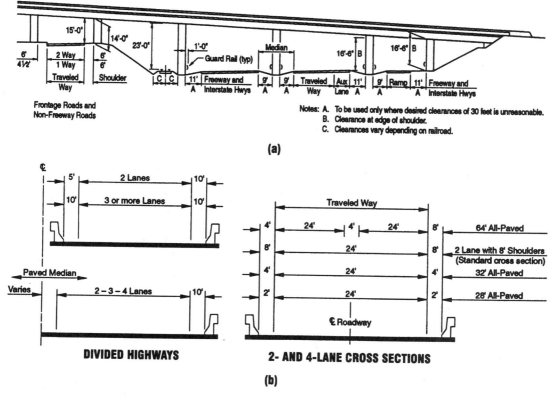

FIGURE A-1 All highway structures have minimum vertical and horizontal clearance requirements. The top figure (a) shows the elevation of a highway bridge with minimum vertical clearances below it. The figure on the bottom (b) shows typical bridge cross-sections and minimum horizontal clearances. Note that long-span bridges may have different details and requirements. *(Merrit)*

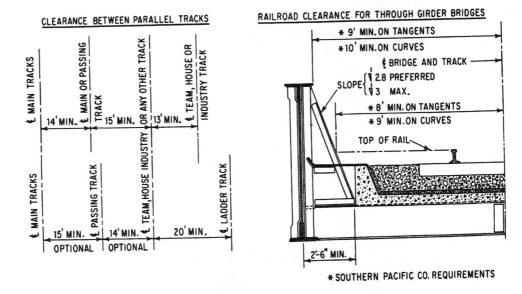

FIGURE A-2 This figure shows the minimum clearances for railroad bridges. Most railroads require that the ballast bed be continuous across bridges to facilitate vertical track adjustments, and that long bridges be equipped with service walkways. *(Merrit)*

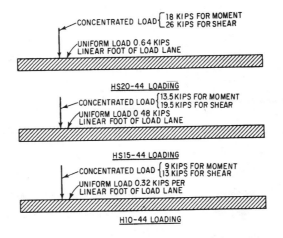

FIGURE A-3 Bridges must support certain loads without exceeding permissible stresses and deflections. This figure shows HS loadings for simply supported spans. For maximum negative moment in continuous spans, an additional concentrated load of equal weight should be placed in one other span for maximum effect. For maximum positive moment, only one concentrated load should be used per lane, but combined with as many spans loaded uniformly as required for maximum effect. *(Merrit)*

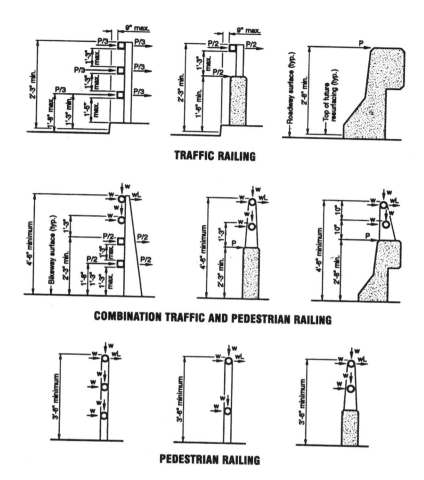

FIGURE A-4 These drawings show service loads for traffic railings, combination traffic and pedestrian railings, and pedestrian railings. $P = 10$ kips, $L =$ post spacing, $w = 50$ lb/ft. Rail loads are shown on the left, and post loads are shown on the right. The rail shapes shown are for illustrative purposes only. *(Merrit)*

One Track of Two Rails

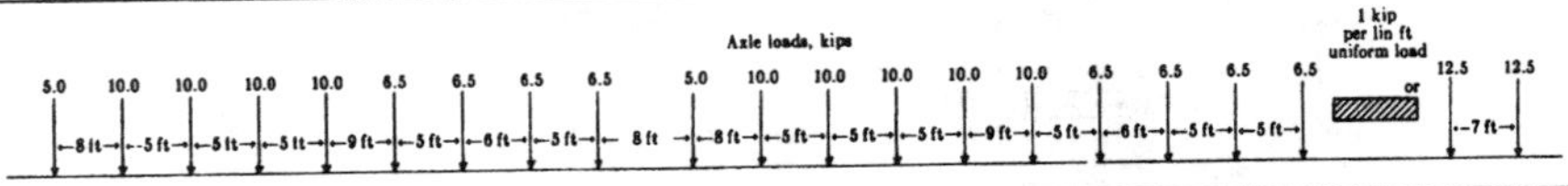

Span, ft	Max moment, ft-kips	Max shear, kips	Max beam reaction, kips[†]	Equivalent uniform load, kips per ft		
				Moment	Shear	Reaction
10	31.2	16.2	20.0	2.50	3.25	2.00
15	62.5	20.0	27.3	2.22	2.67	1.82
20	103.1	25.0	32.8	2.06	2.50	1.64
25	152.5	28.4	37.8	1.95	2.27	1.51
30	205.2	31.5	43.1	1.82	2.10	1.44
35	261.5	34.6	48.8	1.71	1.98	1.39
40	327.8	37.7	54.0	1.64	1.88	1.35
50	475.5	43.5	64.3	1.52	1.74	1.29
60	649.5	48.8	76.6	1.44	1.63	1.28
70	853.7	55.3	88.5	1.39	1.58	1.26
80	1080.0	62.1	99.4	1.35	1.55	1.24
90	1334.7	68.6	109.3	1.32	1.53	1.22
100	1609.7	75.0	118.6	1.29	1.50	1.19
125	2497.7	89.7	140.5	1.28	1.44	1.12
150	3531.0	103.7	162.7	1.25	1.38	1.08
175	4676.3	117.3	185.8	1.22	1.34	1.06
200	5939.0	130.5	209.5	1.19	1.31	1.05
250	8796.3	156.6	257.6	1.13	1.25	1.03

*The standard Class E10 load train consists of two Class E10 engines, coupled front to rear, followed by an indefinite, uniform load of 1 kip per lin ft of track. To obtain the actual design moments, shears, and reactions, the tabulated figures must be multiplied by 8.0 for E80 loading.

[†] From two spans.

FIGURE A-5 This figure is a table showing the maximum moments, shears, and reductions for Class E10 engine loading. *(Merrit)*

Structure Type	Impact, percent*
Prestressed concrete:	
$L < 60$	$35 - \dfrac{L^2}{500} \geq 20$
$L \geq 60$	$\dfrac{800}{L-2} + 14 \geq 20$
Reinforced concrete:	$\dfrac{100LL}{LL+DL}$
Steel:	
Non-steam engine equipment	
$L < 80$	$\dfrac{100}{S} - 40 - \dfrac{3L^2}{1600}$
$L \geq 80$	$\dfrac{100}{S} - 16 + \dfrac{600}{L-30}$
Steam engine equipment	
$L < 100$	$\dfrac{100}{S} - 60 - \dfrac{L^2}{500}$
$L \geq 100$	$\dfrac{100}{S} - 10 + \dfrac{1800}{L-40}$
Truss spans	$\dfrac{100}{S} - 15 + \dfrac{4000}{L+25}$

*For ballasted decks use 90% of calculated impact.
L = span, ft; S = longitudinal beam spacing, ft; DL = applicable dead load; LL = applicable live load.

FIGURE A-6 This figure is a table that shows railroad impact factors for concrete and steel structures. *(Merrit)*

Span, ft	Highway One lane of HS20			Railway One track of E60		
	Live load	Impact	Total	Live load	Impact	Total
50	628	180	808	2,853	1,007	3,860
100	1,524	339	1,863	9,660	2,380	12,040
200	4,100	632	4,732	35,634	6,960	42,594

FIGURE A-7 This figure is a table showing the moments at midspan of highway and railway bridges in ft-kips. *(Merrit)*

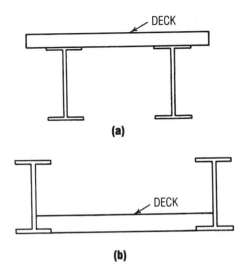

FIGURE A-8 This figure shows deck placements on steel bridges. The top drawing (a) shows a deck-type bridge, where the deck is on top of the main members. The bottom drawing (b) shows a through bridge, where the deck is between the main members. *(Merrit)*

Property	Structural steel	High-strength low-alloy steel		Quenched and tempered low-alloy steel	High-yield-strength, quenched and tempered alloy steel	
AASHTO designation	M270 Grade 36	M270 Grade 50	M270 Grade 50W	M270 Grade 70W	M270 Grades 100/100W	
Equivalent ASTM designations	A709 Grade 36	A709 Grade 50	A709 Grade 50W	A709 Grade 70W	A709 Grades 100/100W	
Thickness of plates, in	Up to 4 incl.	Up to 4 incl.	Up to 4 incl.	Up to 4 incl.	Up to 2.5 incl.	Over 2.5 to 4 incl.
Shapes	All groups	All groups	All groups	Not applicable	Not applicable	Not applicable
Minimum tensile strength, F_u, ksi	58	65	70	90	110	100
Minimum yield point or yield strength, F_y, ksi	36	50	50	70	100	90

FIGURE A-9 This table shows the minimum mechanical properties of structural steel preferred in bridge construction. Properties of the various grades of steel and the testing methods used to control them are regulated by ASTM. *(Merrit)*

Group	γ	D	$(L+I)_n$	$(L+I)_p$	CF	E	B	SF	W	WL	LF	R+S+T	EQ	ICE	% of basic unit stresses
\multicolumn{16}{c}{Service-Load Design†}
I	1.0	1	1	0	1	β_E	1	1	0	0	0	0	0	0	100
IA	1.0	1	2	0	0	0	0	0	0	0	0	0	0	0	150
IB	1.0	1	0	1	1	β_E	1	1	0	0	0	0	0	0	‡
II	1.0	1	0	0	0	1	1	1	1	0	0	0	0	0	125
III	1.0	1	1	0	1	β_E	1	1	0.3	1	1	0	0	0	125
IV	1.0	1	1	0	1	β_E	1	1	0	0	0	1	0	0	125
V	1.0	1	0	0	0	1	1	1	1	0	0	1	0	0	140
VI	1.0	1	1	0	1	β_E	1	1	0.3	1	1	1	0	0	140
VII	1.0	1	0	0	0	1	1	1	0	0	0	0	1	0	133
VIII	1.0	1	1	0	1	1	1	1	0	0	0	0	0	1	140
IX	1.0	1	0	0	0	1	1	1	1	0	0	0	0	1	150
\multicolumn{16}{c}{Load-Factor Design§}
I	1.3	β_D	1.67¶	0	1.0	β_E	1	1	0	0	0	0	0	0	
IA	1.3	β_D	2.20	0	0	0	0	0	0	0	0	0	0	0	
IB	1.3	β_D	0	1	1.0	β_E	1	1	0	0	0	0	0	0	
II	1.3	β_D	0	0	0	β_E	1	1	1	0	0	0	0	0	
III	1.3	β_D	1	0	1	β_E	1	1	0.3	1	1	0	0	0	
IV	1.3	β_D	1	0	1	β_E	1	1	0	0	0	1	0	0	
V	1.25	β_D	0	0	0	β_E	1	1	1	0	0	1	0	0	
VI	1.25	β_D	1	0	1	β_E	1	1	0.3	1	1	1	0	0	
VII	1.3	β_D	0	0	0	β_E	1	1	0	0	0	0	1	0	
VIII	1.3	β_D	1	0	1	β_E	1	1	0	0	0	0	0	1	
IX	1.20	β_D	0	0	0	β_E	1	1	1	0	0	0	0	1	

*D = dead load
L = live load
I = live-load impact
E = earth pressure
B = buoyancy
W = wind load on structure
WL = wind load on live load
LF = longitudinal force from live load
$(L+I)_n$ = live load plus impact for AASHTO highway loading
CF = centrifugal force
F = longitudinal force due to friction
R = rib shortening
S = shrinkage
T = temperature
EQ = earthquake
SF = stream flow pressure
ICE = ice pressure
$(L+I)_p$ = live load plus impact consistent with the overload criteria of the operating agency

† For service-load design: No increase in allowable unit stresses is permitted for members or connections carrying wind loads only.
β_E = 1.0 for lateral loads on rigid frames subjected to full earth pressure
   = 0.5 when positive moment in beams and slabs is reduced by half the earth-pressure moment
Check both loadings to see which one governs.

‡ % = $\dfrac{\text{Maximum unit stress (operating rating)}}{\text{Allowable basic unit stress}} \times 100$

§ For load factor design:
β_E = 1.3 for lateral earth pressure for rigid frames excluding rigid culverts
   = 0.5 for lateral earth pressure when checking positive moments in rigid frames
   = 1.0 for vertical earth pressure
β_D = 0.75 when checking member for minimum axial load and maximum moment or maximum eccentricity and column design
   = 1.0 when checking member for maximum axial load and minimum moment and column design
   = 1.0 for flexural and tension members

¶ β_D = 1.25 for design of outer roadway beam for combination of sidewalk and roadway live load plus impact, if it governs the design, but section capacity should be at least that required for β_D = 1.67 for roadway live load alone
   = 1.00 for deck-slab design for $D+L+I$

FIGURE A-10 This is a table that portrays capacity-reduction and load factors for bridge constructions. *(Merrit)*

	AASHTO	AREA
Min depth-span ratios:		
For rolled beams	1/25	1/15
For girders	1/25	1/12
For trusses	1/10	1/10
Max slenderness ratios:		
For main members in compression	120	100
For bracing members in compression	140	120
For main members in tension	200	200
For bracing members in tension	240	200

FIGURE A-11 This table shows the dimensional limitations for bridge members. *(Merrit)*

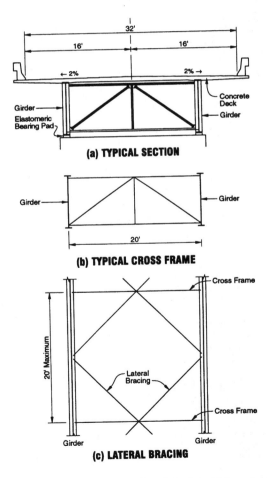

FIGURE A-12 Different types of bracings for a two-lane deck-girder highway bridge are shown here. *(Merrit)*

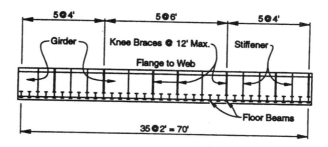

(a) LONGITUDINAL SECTION

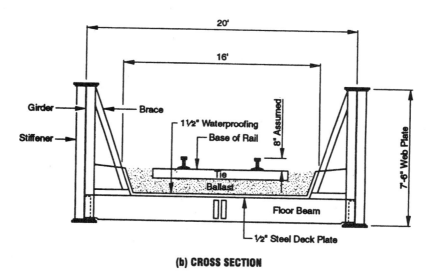

(b) CROSS SECTION

FIGURE A-13 This figure shows two views of a single-track through girder railway bridge with bracing systems. Since top lateral and transverse bracing systems cannot be installed on through-girder spans, the top flanges of the girders must be braced against the floor system (b). *(Merrit)*

(a) Stress Categories for Typical Connections

Type of connection	Figure No.	Stress	Category
Toe of transverse stiffeners	17.13a	Tension or reversal	C
Butt weld at flanges	17.13b	Tension or reversal	B
Gusset for lateral bracing (assumed groove weld, $R \geq 24$ in)	17.13c	Tension or reversal	B
Flange to web	17.13d	Shear	F

(b) Stress Categories for Weld Conditions in Fig. 17.13c

Weld condition*	Category
Unequal thickness; reinforcement in place	E
Unequal thickness; reinforcement removed	D
Equal thickness; reinforcement in place	C
Equal thickness; reinforcement removed	B

(c) Stress Categories for Radii R in Fig. 17.13c

R, in†	Category for welds	
	Fillet	Groove
24 or more	D	B
From 6 to 24	D	C
From 2 to 6	D	D
2 or less	E	E

* For transverse loading, check transition radius for possible assignment of lower category.
† Also applies to transverse loading.

FIGURE A-14 This table shows the fatigue stress categories for bridge members. *(Merrit)*

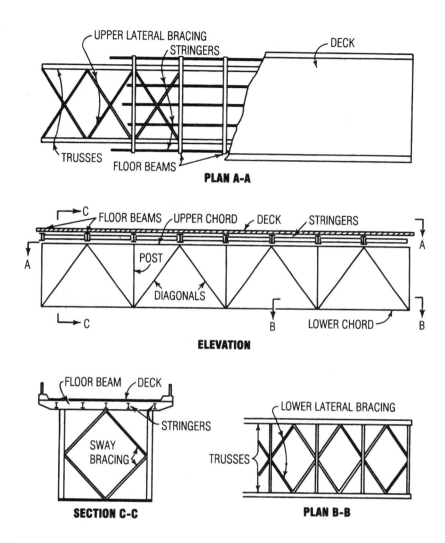

FIGURE A-15 These drawings show bracings on a deck truss bridge. *(Merrit)*

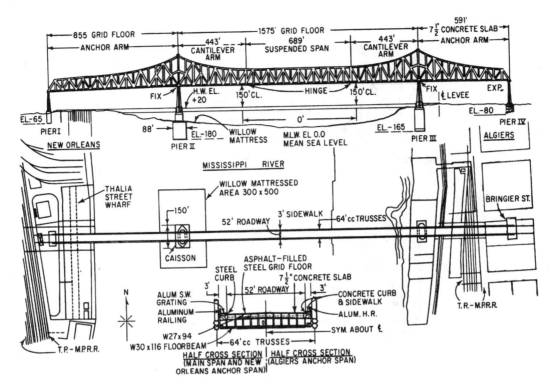

FIGURE A-16 This drawing shows a typical cantilever truss bridge. *(Merrit)*

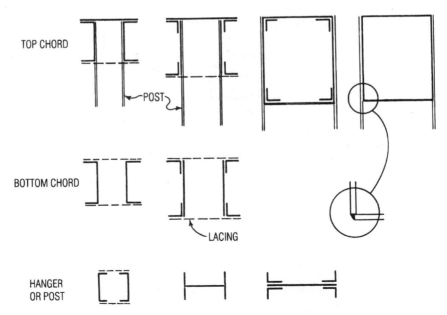

FIGURE A-17 Typical sections that are used in steel bridge trusses are shown in this figure. *(Merrit)*

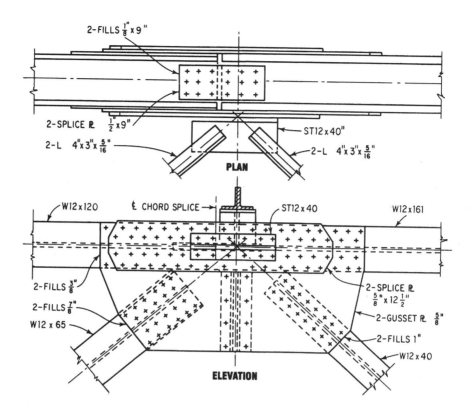

FIGURE A-18 Here, an upper chord joint used in a bridge truss is shown. *(Merrit)*

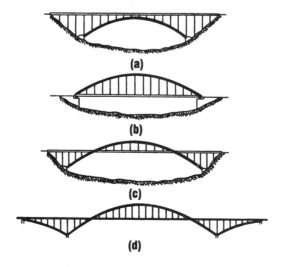

FIGURE A-19 Arch bridges are sometimes preferred for aesthetic reasons. Four basic types of steel arch bridges are shown here: (a) an open spandrel arch, (b) a tied arch, (c) an arch with deck at an intermediate level, and (d) a multiple-arch. *(Merrit)*

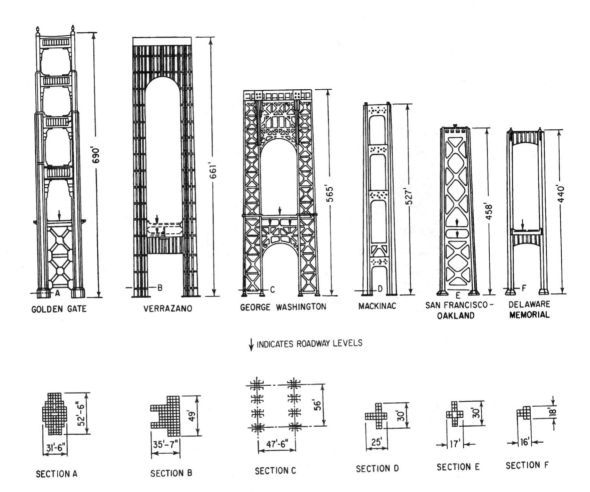

FIGURE A-20 Six different types of towers used for long suspension bridges are shown in these drawings. *(Merrit)*

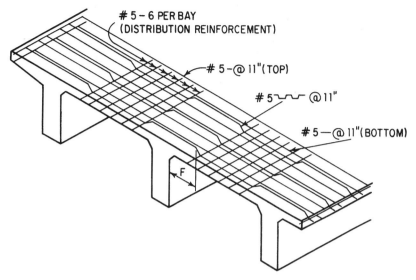

FIGURE A-21 The typical layout of deck reinforcement in a concrete T-beam bridge is shown in this drawing. *(Merrit)*

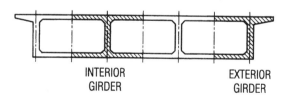

FIGURE A-22 This figure shows the typical design sections (cross-hatched) for a box-girder bridge. *(Merrit)*

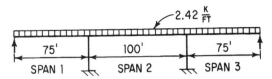
(a) DEAD LOAD ON BRIDGE

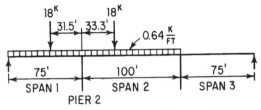
(b) LANE LOADING FOR MAXIMUM MOMENT OVER PIER 2

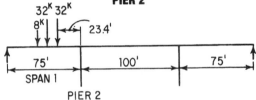
(c) TRUCK LOADING FOR MAXIMUM MOMENT OVER PIER 2

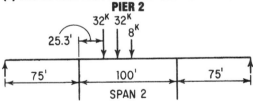
(d) ALTERNATE TRUCK LOADING FOR MAXIMUM MOMENT OVER PIER 2

FIGURE A-23 These drawings show four loading patterns for maximum stresses in a box-girder bridge. *(Merrit)*

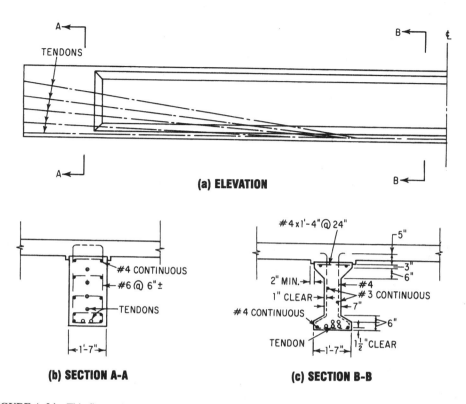

FIGURE A-24 This figure shows a typical precast, prestressed I beam used in highway bridges. *(Merrit)*

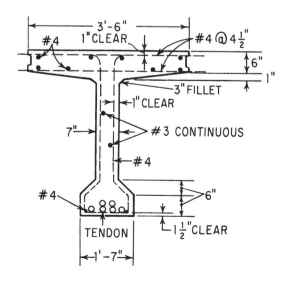

FIGURE A-25 This figure shows a typical precast, prestressed T beam used in highway bridges. *(Merrit)*

APPENDIX B
AIRPORT ENGINEERING

Approach Category (Speed, Knots)	Airplane Design Group (Wing Span, ft)					
	I Less than 49	II 49 to 78	III 79 to 117	IV 118 to 170	V 171 to 213	VI 214 to 261
A (less than 90)	A-I*	A-II*,†	A-III†	A-IV†		
B (91 to 120)	B-I*,†	B-II*,†	B-III†	B-IV†,‡		
C (121 to 140)	C-I†	C-II*,†	C-III†	C-IV†,‡	C-V‡	C-VI‡
D (141 to 165)	D-I†	D-II†	D-III†,‡	D-IV†,‡	D-V‡	
E (166 or more)		E-II†	E-III†,‡			

*Small airplanes (12,500 lb or less maximum takeoff weight). Examples:
- A-I: Cessna 177 Cardinal
- A-II: DHC-6 Twin Otter
- B-I: Beech 100 King Air
- B-II: Beech 200 Super King Air
- C-II: Rockwell 980

† Large airplanes (more than 12,500 lb maximum takeoff weight). Examples:
- A-II: Dassault 941
- A-III: DHC-8 Dash 8-300
- A-IV: Lockheed 1649 Constellation
- B-I: Mitsubishi 300 Diamond
- B-II: Cessna III Citation
- B-III: BAe 146
- B-IV: MDC DC-7
- C-I: Gates 55 Learjet
- C-II: Gruman III Gulfstream
- C-III: Boeing 737 500
- C-IV: Boeing 757
- D-I: Gates 36A Learjet
- D-II: Gruman IV Gulfstream
- D-III: BAC III 500
- D-IV: Boeing 707-200
- E-II: Lockheed SR-71 Blackbird

‡ Heavy airplanes (takeoff weight of 300,000 lb or more). Examples:
- B-IV: Ilyushin Il-76
- C-IV: Airbus A-300-B4
- C-V: Boeing 747-SP
- C-VI: Lockheed C-5B Galaxy
- D-III: BAC\Aerospatiale Concord
- D-IV: Boeing 777
- D-V: Boeing 747-400
- E-III: Tupoleu TU-144

FIGURE B-1 This table shows the airplane operational characteristics for the FAA Airport Reference Coding System. *(Merrit)*

APPENDIX B

Item		Airports Serving Aircraft Approach Categories[b] A and B					Airports Serving Aircraft Approach Categories[b] C and D					
		Airplane Design Group[c]					Airplane Design Group[c]					
		I[d] ft	I ft	II ft	III ft	IV ft	I ft	II ft	III ft	IV ft	V ft	VI ft
Length, ft:												
Runway[e]		2,800	3,200	4,370	5,360	6,370	5,490	6,370	7,290	9,580	10,700	12,000
Runway safety area (beyond runway end)	PR[f]	600	600	600	800	1,000	1,000	1,000	1,000	1,000	1,000	1,000
	NP, V[g]	240	240	300	600	1,000						
Runway object-free area (beyond runway end)	PR	300	500	600	1,000	1,000	1,000	1,000	1,000	1,000	1,000	1,000
	NP, V	1,000	1,000	1,000	1,000	1,000						
Width, ft:												
Runway	PR	75	100	100	100	150	100	100	100	150	150	200
	NP, V	60	60	75	100	150						
Runway safety area	PR	300	300	300	400	500	500	500	500	500	500	500
	NP, V	120	120	150	300	500						
Runway object-free area	PR	800	800	800	800	800	800	800	800	800	800	800
	NP, V	250	400	500	800	800						
Taxiway		25	25	35	50	75	25	35	50	75	75	100
Taxiway safety area		49	49	79	118	171	49	79	118	171	214	262
Taxiway object-free area		89	89	131	186	259	89	131	186	259	320	386
Taxilane object-free area		79	79	115	162	225	79	115	162	225	276	334
Minimum distance between:												
Center lines of parallel runways[b]	PR	2,500	2,500	2,500	2,500	2,500	2,500	2,500	2,500	2,500	2,500	2,500
	V	700	700	700	700	700	700	700	700	700	1,200	1,200
Center lines of runway and center line of taxiway	PR	200	250	300	350	400	400	400	400	400	500	600
	NP, V	150	225	240	300	400	300	300	400	400	500	600
Center line of runway and aircraft parking area	PR	400	400	400	400	500	500	500	500	500	500	500
	NP, V	125	200	250	400	50	400	400	500	500	500	500
Center line of taxiway and aircraft parking apron		45	45	66	93	130	45	66	93	130	160	193
Center line of parallel taxiways		69	69	105	152	215	69	105	152	215	267	324
Center line of runway to building line or obstruction[i]	PR	875	875	875	875	875	875	875	875	875	875	875
	NP, V	600	600	600	600	600	713	713	713	713	713	713
Centerline of taxiway to obstruction		45	45	66	93	130	45	66	93	130	160	193
Maximum runway grades, %[j]:												
Longitudinal		2.0	2.0	2.0	2.0	2.0	1.5	1.5	1.5	1.5	1.5	1.5
Transverse[k]		2.0	2.0	2.0	2.0	2.0	1.5	1.5	1.5	1.5	1.5	1.5

[a] "Airport Design," FAA Advisory Circular 150/5300-13.
[b] Aircraft Approach Categories are described in Art. 18.3.
[c] Airplane Design Group is described in Art. 18.3.
[d] Represents airports serving only small airplanes (an airplane of 12,500 lb or less maximum certificated takeoff weight).
[e] Runway lengths assume an airport elevation of 1000 ft above mean sea level (MSL) and a mean daily maximum temperature of 85° in the hottest month. Actual runway lengths should be based on the selected design airplane adjusted for the local conditions of elevation, temperature, and runway gradient. The lengths shown are representative of a runway that can accommodate selected airplanes found in the indicated Airport Reference Code (ARC). Runway length for airplanes over 60,000 lb is usually determined based on the amount of fuel needed to fly a certain distance or haul length and may need to be increased from that determined above.
[f] Values are for runways serving precision instrument approach system (PR).
[g] Values are for runways serving visual approaches (V) or nonprecision instrument approaches (NP).

FIGURE B-2 This table summarizes physical characteristics for airport design set by national standards. These are the minimum requirements that the FAA considers acceptable for safe operation. *(Merrit)*

NPIAS[†] airport category	Classification description
Commercial service	Annual scheduled passenger service
Primary	Percent of total U.S. enplanements
Large hub (L)	1% or more
Medium hub (M)	0.25 – 1.00%
Small hub (S)	0.05 – 0.25%
Non-hub (N)	0.01 – 0.05%
Other	2500 or more but less than 0.01%
Reliever[‡]	Must have at least:
	50 based aircraft, or
	25,000 annual itinerant operations, or
	35,000 annual local operations
General aviation	All other civil airports

[*] Established by the Federal Aviation Administration.
[†] National Plan of Integrated Airport Systems.
[‡] Intended to reduce congestion at large commercial service airports by providing general aviation pilots with alternative landing areas and by providing more general aviation access to the overall community.

FIGURE B-3 This table shows the aeronautical activity levels for the Functional-Role Airport Classification System, established by the FAA. *(Merrit)*

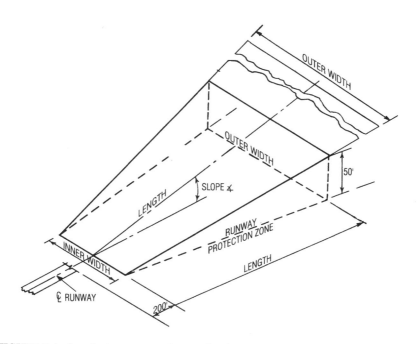

FIGURE B-4 In order to test approach zones for clearance of obstructions, a topographic map of the airport site and its environs is required. This figure shows the runway protection zones and approach surfaces. *(Courtesy of FAA)*

Dimension	Item	Dimensional standards, ft (see Fig. 18.1)					
		Visual runway		Nonprecision instrument runway			Precision instrument runway
		Utility runways[‡]	Runways larger than utility	Utility runways	Runways larger than utility		
					Visibility minimums greater than ¾ mi	Visibility minimums as low as ¾ mi	
A	Width of primary surface and width of approach surface at inner end	250	500	500	500	1,000	1,000
B	Radius of horizontal surface	5,000	5,000	5,000	10,000	10,000	10,000
C	Approach surface width at end	1,250	1,500	2,000	3,500	4,000	16,000
D	Approach surface length	5,000	5,000	5,000	10,000	10,000	[†]
E	Approach slope	20:1	20:1	20:1	34:1	34:1	[†]

[*] Federal Aviation Administration.
[†] Precision instrument approach slope is 50:1 for inner 10,000 ft and 40:1 for an additional 40,000 ft.
[‡] Runways expected to serve propeller-driven airplanes with maximum certificated takeoff weight of 12,500 lb or less.

FIGURE B-5 This table shows the criteria for airport imaginary surfaces. The FAA has established standards which set up civil imaginary surfaces for determining obstructions to airports. Objects that extend above these surfaces are considered obstructions and should be removed or marked and lighted. *(Merrit)*

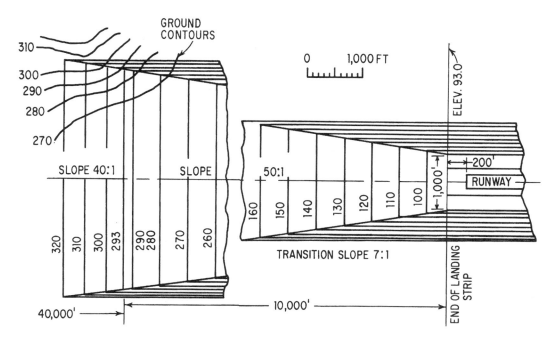

FIGURE B-6 This figure shows a template used for checking approach-zone clearance for instrument runways. The template is used to compare the ground-surface contours with those of the runway approach surfaces so that the runway layout can be adjusted, if necessary, to avoid obstacles with a minimum sacrifice of wind coverage. *(Merrit)*

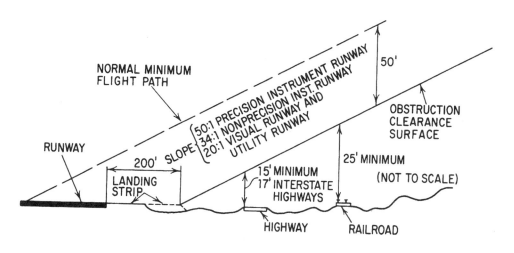

FIGURE B-7 This drawing illustrates the vertical profile along the extended center line of the runway that shows the minimum clearance required by the FAA over highways and railways. *(Merrit)*

Facilities expected to serve	Runway end		(a) Approach surface dimensions				(b) Runway Protection Zone (RPZ) Dimensions for Approach End			
	Approach end	Opposite end	Length, ft	Inner width, ft	Outer width, ft	Slope, run/rise	Length, ft	Inner width, ft	Outer width, ft	RPZ, acres
Only small airplanes	V	V	5,000	250	1,250	20:1	1,000	250	450	8.035
		NP	5,000	500	1,250	20:1	1,000	500	650	13.200
		NP ¾ P	5,000	1,000	1,250	20:1	1,000	1,000	1,050	23.542
	NP	V	5,000	500	2,000	20:1	1,000	500	800	14.922
		NP NP ¾ P	5,000	1,000	2,000	20:1	1,000	1,000	1,200	25.252
Large airplanes	V	V	5,000	500	1,500	20:1	1,000	500	700	13.770
		NP ¾ P	5,000	1,000	1,500	20:1	1,000	1,000	1,100	24.105
	NP	V	10,000	500	3,500	34:1	1,700	500	1,010	29.465
		NP NP ¾ P	10,000	1,000	3,500	34:1	1,700	1,000	1,425	47.320
All airplanes	NP ¾	V	10,000	1,000	4,000	34:1	1,700	1,000	1,510	48.978
		NP NP ¾ P								
	P	V	10,000 plus	1,000	4,000	50:1	2,500	1,000	1,750	78.914
		NP NP ¾ P	40,000	4,000	16,000	40:1				

Small airplane: Airplanes of 12,500 lb or less maximum certificated takeoff weight
Large airplane: Airplanes of more than 12,500 lb maximum certificated takeoff weight
V—visual approach
NP—nonprecision instrument approach with visibility minimums more than ¾ statute mile
NP ¾—nonprecision instrument approach with visibility minimums as low as ¾ statute mile
P—precision instrument approach

FIGURE B-8 This is a table showing the clearance dimensions for airport approaches. *(Merrit)*

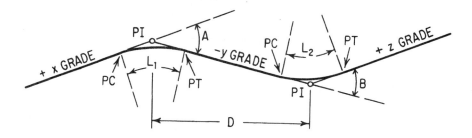

FIGURE B-9 This drawing shows the vertical profile around the runway center line that illustrates changes in longitudinal grades. Longitudinal changes should be avoided. If changes are necessary, they should be in accordance with the table in Figure B-10. *(Courtesy of FAA)*

	Runways serving categories A and B airplanes	Runways serving categories C and D airplanes
Maximum gradient at ends of runway, such as x grade or z grade (Fig. 18.5)	0 to 2.0%	0 to 0.8%, first and last quarter of runway length
Maximum gradient in middle portion of runway, such as y grade (Fig. 18.5)	0 to 2.0%	0 to 1.5%
Maximum grade change, such as A or B (Fig. 18.5)	2.0%	1.5%
Minimum length of vertical curve L_1 or L_2 (Fig. 18.5) for each 1.0% grade change	300 ft*	1000 ft
Minimum distance between points of intersection for vertical curves, D (fig. 18.5)	$250(A + B)$ ft†	$1000(A + B)$ ft†

* Vertical curves not required at utility airports for grade changes less than 0.4%.
† A% and B% are successive changes in grade.

FIGURE B-10 This table outlines vertical curve data and maximum grade changes for runways. *(Merrit)*

Aircraft Reference Code	Cross-wind component, knots
A-I and B-I	10.5
A-II and B-II	13
A-III and B-III	16
C-I through D-III	16
A-IV through D-VI	20

FIGURE B-11 Because light airplanes are more susceptible to cross winds than heavy airplanes, allowable cross-wind components are specified for runways designed to serve aircraft of different Aircraft Reference Codes. This table shows the allowable cross-wind components for aircraft. *(Merrit)*

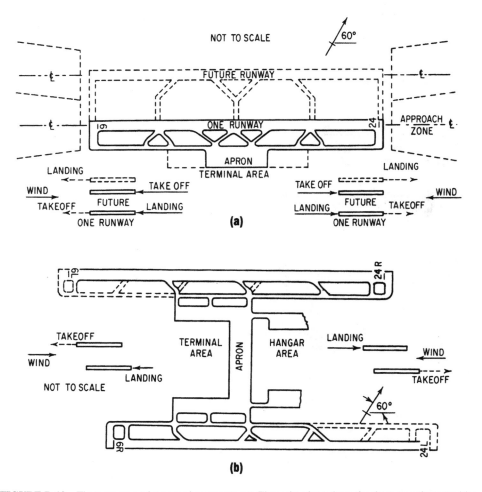

FIGURE B-12 There are several runway layout patterns. These drawings show simple runway layouts: (a) a single runway with a future parallel runway, and (b) two parallel runways. *(Merrit)*

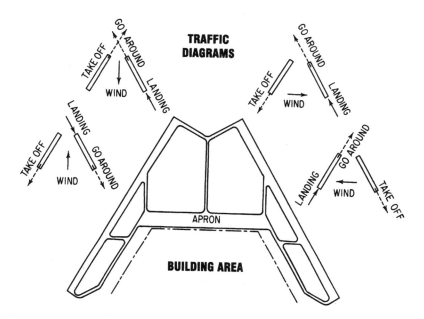

FIGURE B-13 This figure shows a V-type runway layout that permits two-directional operation of aircraft. *(Courtesy of FAA)*

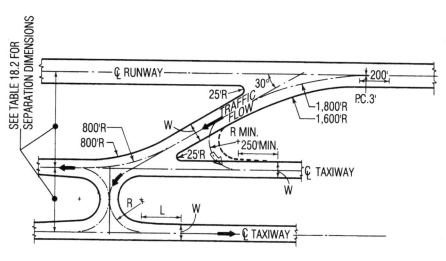

FIGURE B-14 An angled-exit taxiway design with dual parallel and crossover is shown here. *(Courtesy of FAA)*

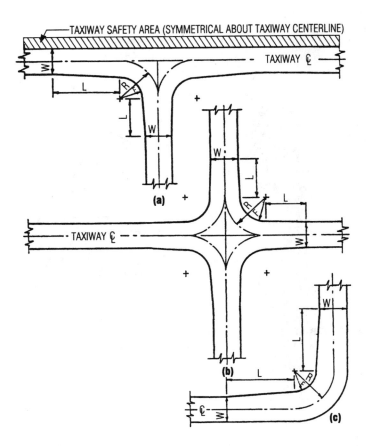

FIGURE B-15 These drawings show taxiway intersection details: (a) a T-shape intersection, (b) a crossover, and (c) a turn. The dimensions *W, R,* and *L* are shown in the following table (Figure B-16). Note that the taxiway safety area shown in the top drawing has been omitted from the other drawings for clarity. *(Merrit)*

		Airplane Design Group					
Design item	Symbol[a]	I	II	III	IV	V	VI
Taxiway width	W	25	35	50[b]	75	75	100
Taxiway edge safety margin[c]		5	7.5	10[d]	15	15	20
Taxiway pavement fillet configuration:							
Radius of taxiway turn[e]	R	75	75	100[f]	150	150	170
Length of lead-in to fillet	L	50	50	150[f]	250	250	250
Fillet radius for center line	F	60	55	55[f]	85	85	85
Fillet radius for judgmental Oversteering, symmetrical widening[g]	F	62.5	57.5	68[f]	105	105	110
Fillet radius for judgmental oversteering, symmetrical widening[h]	F	62.5	57.5	60[f]	97	97	100
Taxiway shoulder width		10	10	20	25	35[i]	40[i]
Taxiway safety area width		49	79	118	171	214	262
Taxiway object-free area width		89	131	186	259	320	386
Taxilane object-free area width		79	115	162	225	276	334

[a] Letters correspond to the dimensions in Fig. 18.10.
[b] For airplanes in Airplane Design Group III with wheelbase equal to or greater than 60 ft, the standard taxiway width is 60 ft.
[c] The taxiway edge safety margin is the minimum acceptable distance between the outside of the airplane wheels and the pavement edge.
[d] For airplanes in Airplane Design Group III with a wheelbase equal to or greater than 60 ft, the taxiway edge safety margin is 15 ft.
[e] Dimensions for taxiway fillet designs relate to the radius of taxiway turn specified. Additional design data can be found in "Airport Design," AC 150/5200-13.
[f] Airplanes in Airplane Design Group II with a wheelbase equal to or greater than 60 ft should use a fillet radius of 50 ft.
[g] Figure 18.10b displays pavement fillets with symmetrical taxiway widening.
[h] Figure 18.10c displays a pavement fillet with taxiway widening on one side.
[i] Airplanes in Airplane Design Groups V and VI normally require stabilized or paved taxiway shoulder surfaces.

FIGURE B-16 This table illustrates dimensional standards for taxiways. *(Merrit)*

Area	Spacing	Depth
Runways and taxiways	Along center line, 200 ft c to c	Cut areas: 10 ft below finished grade Fill areas: 10 ft below existing ground surface*
Other areas of pavement	One boring per 10,000 ft² of area	Cut areas: 10 ft below finished grade Fill areas: 10 ft below existing ground surface*
Borrow areas	Sufficient tests to define borrow material clearly	To depth of proposed excavation of borrow

*For deep fills, boring depths should be used as necessary to determine the extent of consolidation and slippage that the fill to be placed may cause.

FIGURE B-17 The FAA has adopted the Unified System of Soil Classification. This table lists the recommended spacing and depths for borings for soil investigations for airport construction. *(Merrit)*

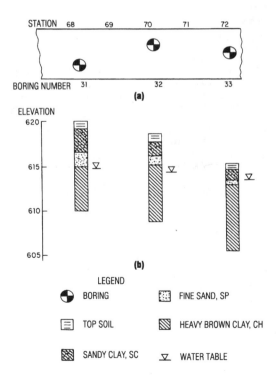

FIGURE B-18 This figure shows borings for a subgrade investigation. The top drawing (a) is a plan of a runway showing the locations of borings, and the bottom drawing (b) shows a typical graphic boring log. *(Courtesy of FAA)*

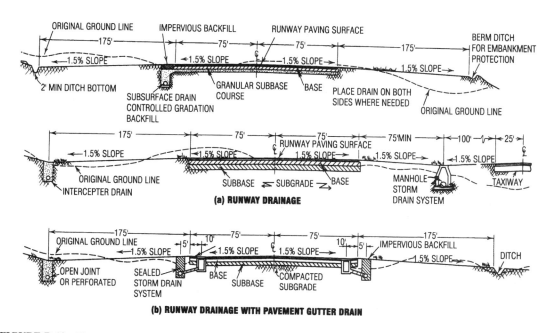

FIGURE B-19 These runway cross sections show typical provisions for drainage. The top drawing (a) shows runoff drainage into shallow ditches, and the bottom drawing (b) shows runoff drainage into a gutter. *(Merrit)*

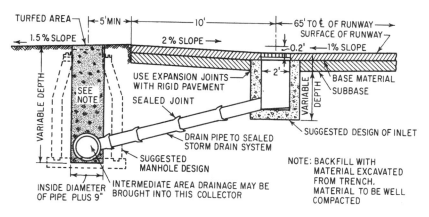

FIGURE B-20 In northern climates, a drainage inlet may be placed at the outer edge of the runway, as shown here. *(Courtesy of FAA)*

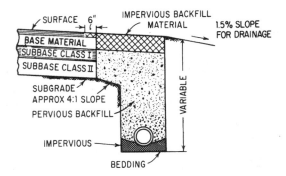

FIGURE B-21 This drawing shows a combined interceptor and base drain. *(Courtesy of FAA)*

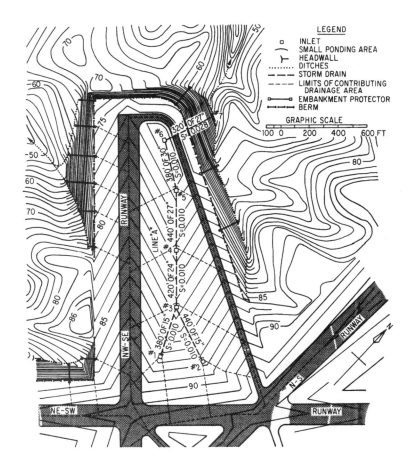

FIGURE B-22 This figure shows the plan of a portion of an airport drainage system. *(Courtesy of FAA Advisory Circular AC 150/5320-5)*

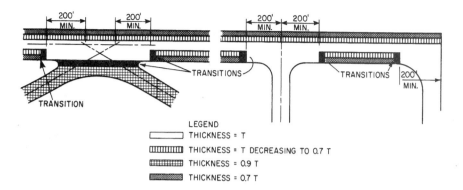

FIGURE B-23 The critical areas—areas subject to the most adverse aircraft loadings—of airport paving are shown here. T = total thickness of flexible pavement or concrete thickness for rigid pavement. See also Figure B-24. *(Merrit)*

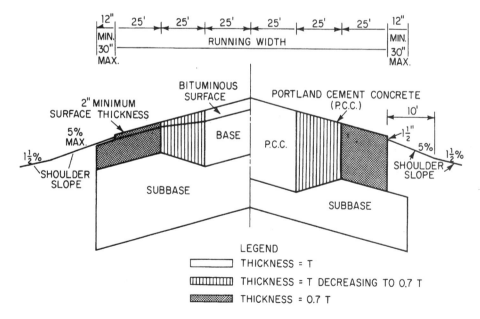

FIGURE B-24 This is a cross section showing typical bituminous pavement (left of center line) and portland cement-concrete pavement construction (right of center line) for critical areas of runways. *(Merrit)*

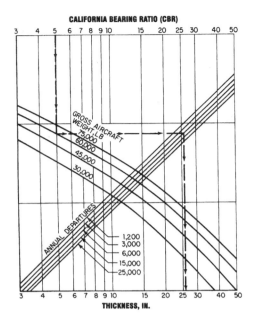

FIGURE B-25 Flexible-pavement design curves for critical areas of single wheel gear are shown here. *(Courtesy of "Airport Pavement," FAA)*

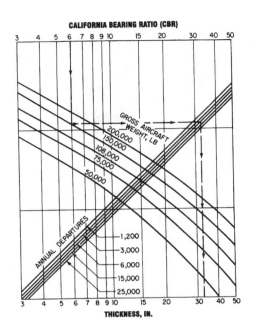

FIGURE B-26 This figure shows flexible-pavement design curves for critical areas of dual-wheel gear. *(Courtesy of "Airport Pavement," FAA)*

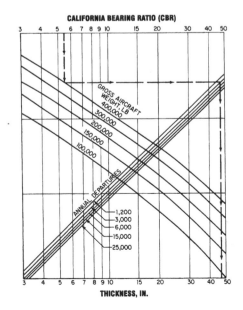

FIGURE B-27 In this drawing, flexible-pavement design curves for critical areas of dual-tandem gear are shown. *(Courtesy of "Airport Pavement," FAA)*

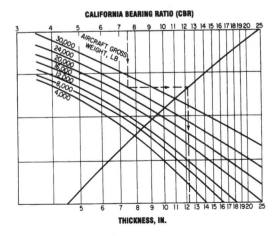

FIGURE B-28 Here, flexible-pavement design curves for light aircraft are shown. A minimum of two inches is required for the surface course. *(Courtesy of "Airport Pavement," FAA)*

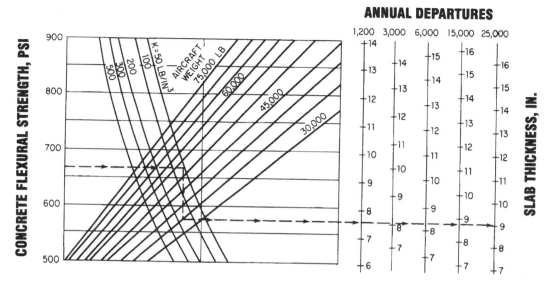

FIGURE B-29 In this figure, rigid-pavement design curves of single-wheel gear are shown. *(Courtesy of "Airport Pavement," FAA)*

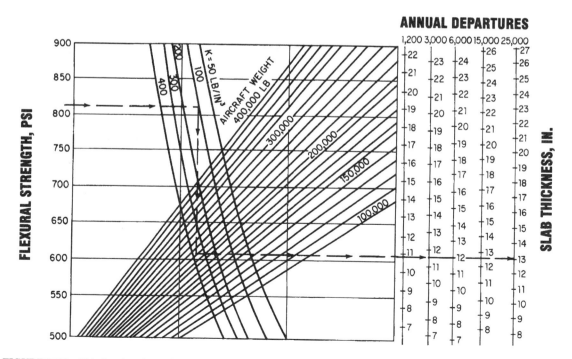

FIGURE B-30 This drawing shows rigid-pavement design curves for dual-wheel gear. *(Courtesy of "Airport Pavement," FAA)*

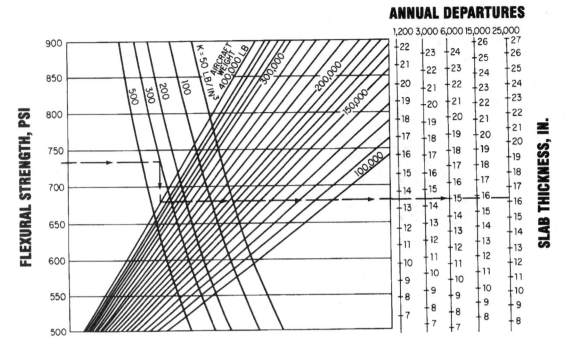

FIGURE B-31 Rigid-pavement design curves for dual-tandem gear is shown in this figure. *(Courtesy of "Airport Pavement," FAA)*

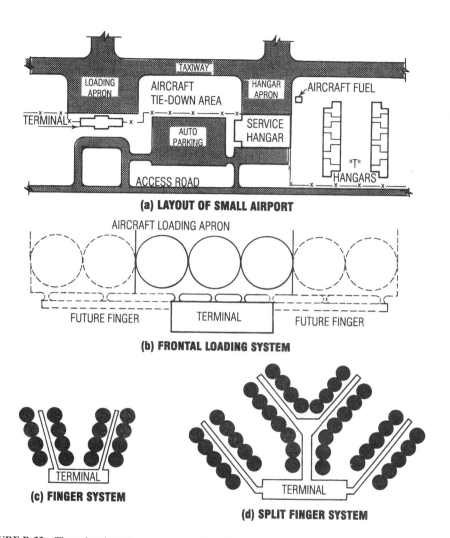

FIGURE B-32 These drawings show the layout of simple terminal systems of low activity. Fingers, shown in (c) and (d), are added to increase aircraft parking capacity. *(Merrit)*

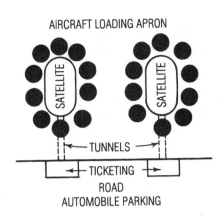

**(a) SATELLITE SYSTEM
(LOS ANGELES INTERNATIONAL AIRPORT)**

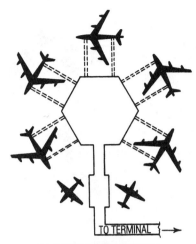

**(b) FINGER-PIER SYSTEM
(U.A.L.-SAN FRANCISCO INTERNATIONAL AIRPORT)**

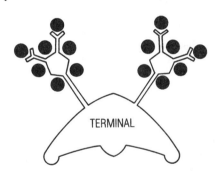

**(c) FINGER-PIER SYSTEM
(T.W.A.-J.F. KENNEDY INTERNATIONAL AIRPORT)**

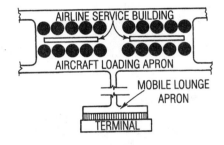

**(d) REMOTE PARKING SYSTEM
(DULLES INTERNATIONAL AIRPORT)**

FIGURE B-33 These drawings show terminal systems that are used at some larger, international airports. *(Merrit)*

(a) MOBILE LOUNGE TERMINAL CONNECTION (FEDERAL AVIATION AGENCY)

(b) MOBILE LOUNGE AIRCRAFT CONNECTION (FEDERAL AVIATION AGENCY)

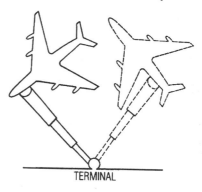

(c) TELESCOPING GANGPLANK PARALLEL PARKING

(d) TELESCOPING GANGPLANK ANGLE PARKING

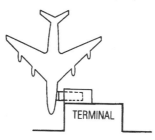
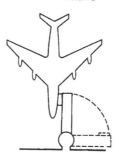

(e) PEDESTAL-TYPE NOSE LOADING DEVICE

(f) PIVOT-TYPE NOSE LOADING DEVICE

FIGURE B-34 Different types of passenger-transfer devices for loading and unloading aircraft at airports are shown in these drawings. *(Merrit)*

Name and model	Gross weight, lb	Wing span	Length	Height
Single Engine, Prop.				
Beech Bonanza	3,125	33 ft 5 in	25 ft 2 in	7 ft 7 in
Cessna 210	2,900	36 ft 7 in	27 ft 9 in	8 ft 8 in
Piper Saratoga	2,900	36 ft 0 in	24 ft 11 in	7 ft 3 in
Multiengine, Prop.				
Aero Commander	8,000	49 ft 0 in	35 ft 1 in	14 ft 6 in
Beech Super King Air	12,500	54 ft 6 in	43 ft 10 in	15 ft 0 in
Cessna Conquest	9,925	49 ft 4 in	39 ft 0 in	13 ft 1 in
Piper Cheyenne	12,050	47 ft 8 in	43 ft 5 in	17 ft 0 in
Executive Jets				
Lockheed Jetstar	35,000	54 ft 5 in	60 ft 5 in	20 ft 5 in
Grumman Gulfstream II	51,340	68 ft 10 in	79 ft 11 in	24 ft 6 in
Learjet 25	13,300	35 ft 7 in	47 ft 7 in	12 ft 7 in
Rockwell Sabreliner	17,500	44 ft 5 in	43 ft 9 in	16 ft 0 in
Airline Transports				
Airbus A-300	330,700	147 ft 1 in	175 ft 6 in	55 ft 6 in
B-737-200	100,800	93 ft 0 in	100 ft 0 in	36 ft 9 in
B-727-200	173,000	108 ft 0 in	153 ft 2 in	34 ft 0 in
DC-9-30	109,000	93 ft 4 in	107 ft 0 in	27 ft 6 in
DC-8-63	358,000	148 ft 5 in	187 ft 5 in	43 ft 0 in
B-747	775,000	195 ft 8 in	229 ft 2 in	64 ft 8 in
B-757	225,000	124 ft 10 in	155 ft 4 in	45 ft 1 in
B-767	350,000	156 ft 1 in	180 ft 4 in	52 ft 7 in
L-1011	432,000	155 ft 4 in	178 ft 8 in	55 ft 10 in
DC-10-30	555,000	161 ft 4 in	181 ft 11 in	59 ft 7 in
MOC-MD-11	602,500	169 ft 10 in	201 ft 4 in	57 ft 10 in

FIGURE B-35 This table lists the physical data—weight, wing span, length, and height—for selected aircraft. *(Merrit)*

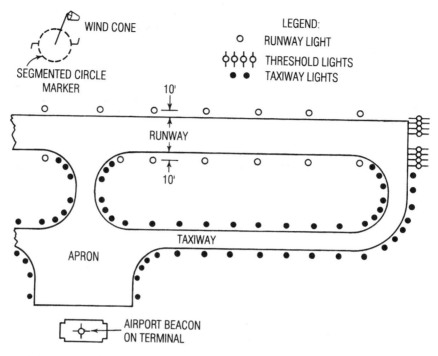

FIGURE B-36 Runway lighting is used to outline the edges of paved runways or to define unpaved runways. This drawing shows a basic layout for airport lighting. *(Merrit)*

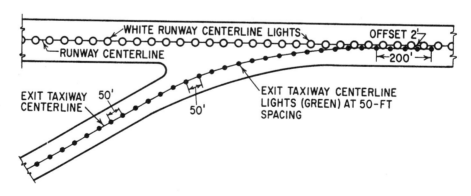

FIGURE B-37 Long-radius taxiway-turnoff lighting is shown here. A longitudinal tolerance may be necessary to avoid joints in rigid pavements. *(Courtesy of FAA)*

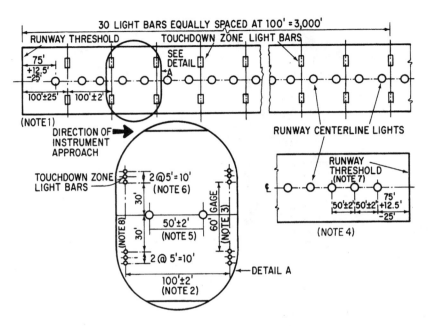

FIGURE B-38 This drawing shows the lighting layout for the runway touchdown zone and center-line runway lights. NOTES: (1) In cased of unusual joint location in concrete pavement, the first pair of light bars may be located 75 to 125 ft. from the threshold. (2) Longitudinal tolerance should not exceed 2 ft. (3) Gage may be reduced to 55 ft. to meet construction requirements. (4) Longitudinal installation tolerance for individual lights should not exceed 2 ft. (5) Center-line lights need not be aligned with transverse light bars. (6) Maximum uniform spacing of lights is 5 ft. to c. (7) Center-line lights may be located up to 2 ft. from the runway center line to avoid joints. (8) Corresponding pairs of transverse light bars should lie along a line perpendicular to the runway center line. *(Courtesy of FAA)*

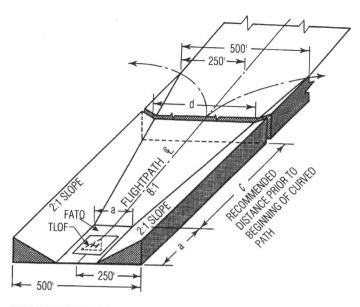

FIGURE B-39 This figure shows the standard dimensions for heliports and approaches. FATO length and width *a* should be at least 1.5 times the overall length of the design helicopter. Straightway approach-departure length *c*, width *d* of the flight path at the wide end of the straightway, and the radius of the curved path should each be at least 300 ft. *(Merrit)*

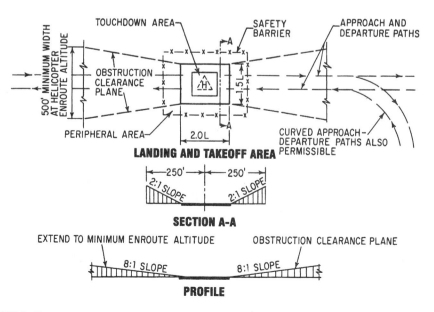

FIGURE B-40 This drawing shows the layout for a small heliport. *L*= overall length of design helicopter. *(Courtesy of "Heliport Design," FAA)*.

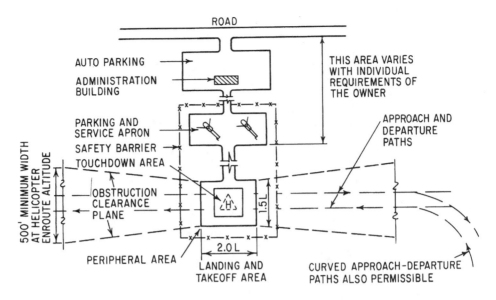

FIGURE B-41 Here, the layout for a large heliport is shown. L = overall length of design helicopter. *(Courtesy of FAA)*

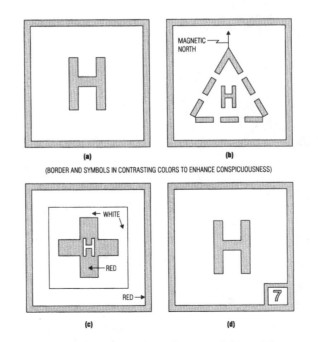

FIGURE B-42 Standard heliport markers, placed near the center of the touchdown area, are shown here: (a) a standard public heliport marker, (b) an example of a marker for private-use ports, (c) a marker for a hospital heliport, and (d) a weight-limiting marker (7000 lb. indicated) for elevated heliports. *(Courtesy of "Heliport Design," FAA)*

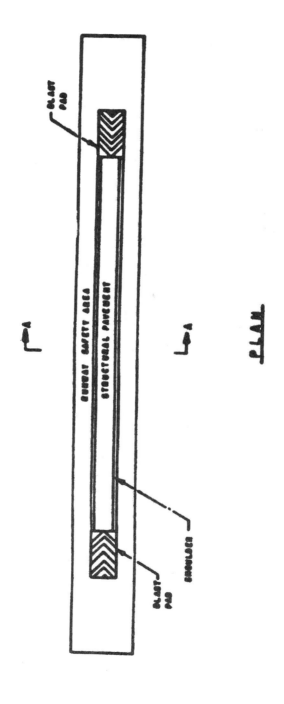

FIGURE B-43 Runway safety area. (*U.S. Department of Transportation, Federal Aviation Administration*)

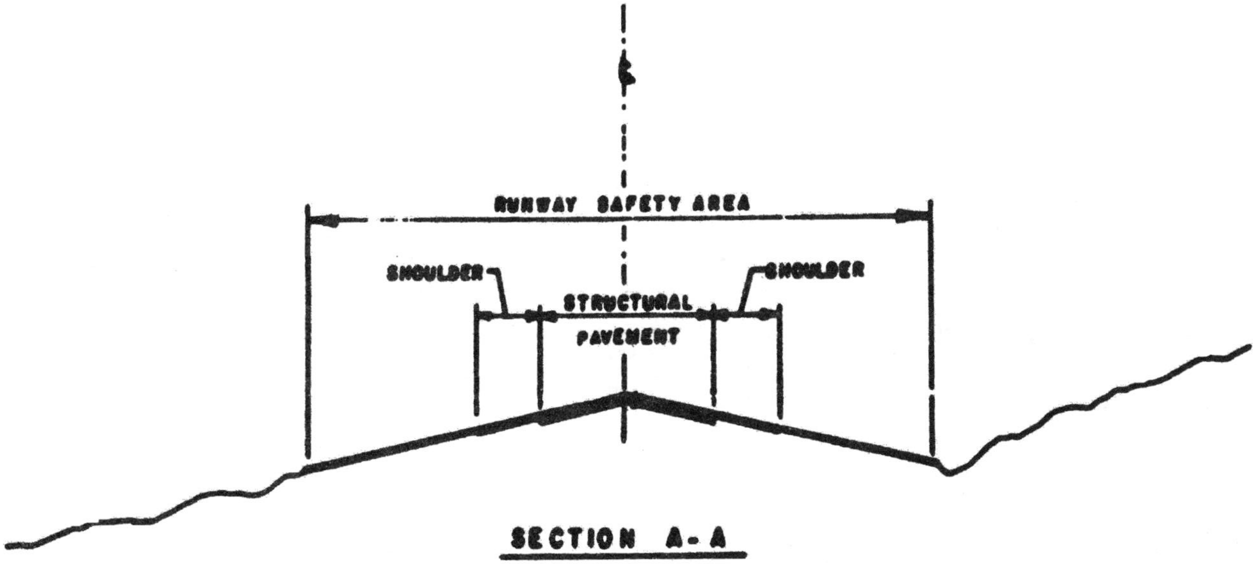

FIGURE B-44 (continued) Runway safety area. (*U.S. Department of Transportation, Federal Aviation Administration*)

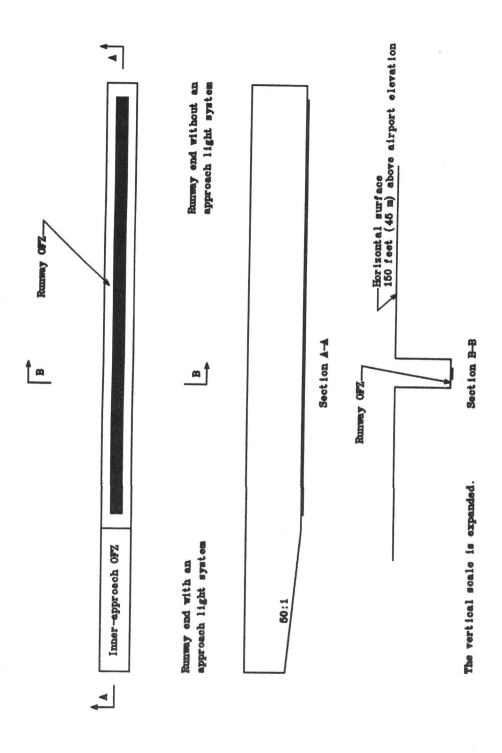

FIGURE B-45 Obstacle free zone (OFZ) for visual runways and runways with not lower than 3/4-statute mile (1,200m) approach visibility minimums. (*U.S. Department of Transportation, Federal Aviation Administration*)

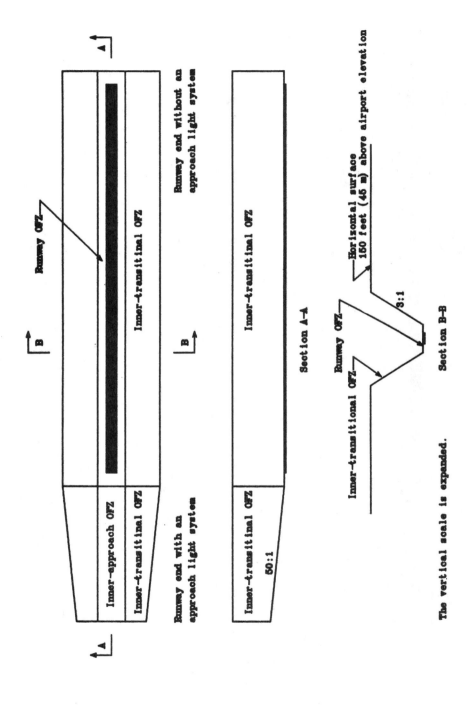

FIGURE B-46 Obstacle free zone (OFZ) for runways serving small airplanes exclusively with lower than 3/4-statute mile (1,200 m) approach visibility minimums. (*U.S. Department of Transportation, Federal Aviation Administration*)

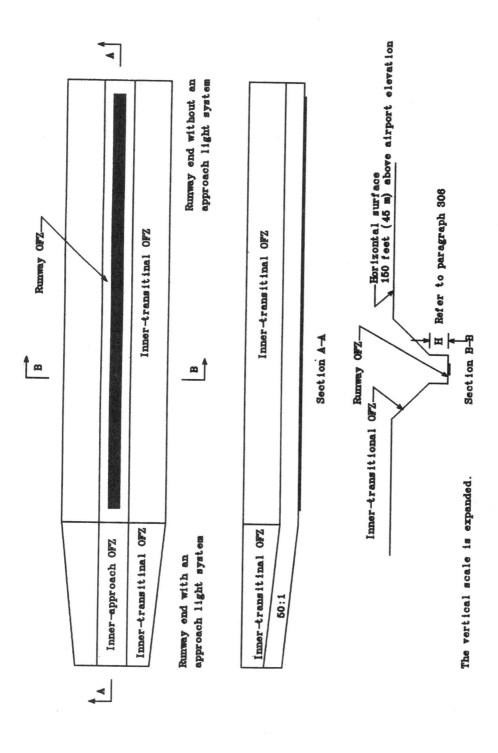

FIGURE B-47 Obstacle free zone (OFZ) for runways serving large airplanes with lower than 3/4-statute mile (1,200 m) approach visibility minimums. (*U.S. Department of Transportation, Federal Aviation Administration*)

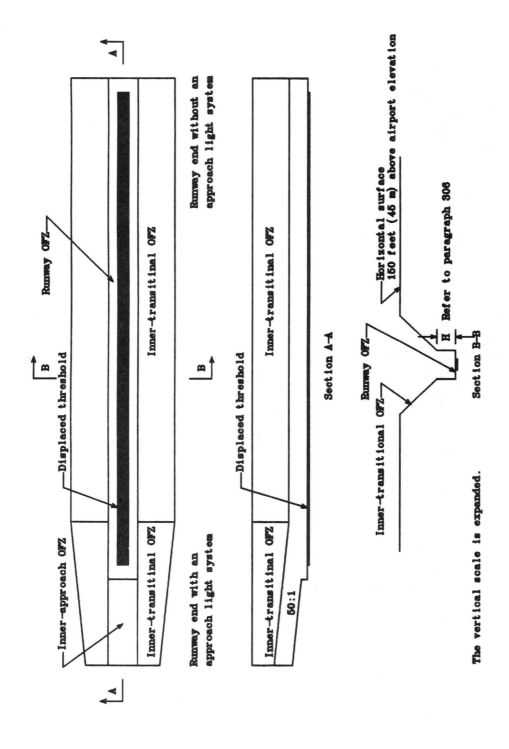

FIGURE B-48 Obstacle free zone (OFZ) for runways serving large airplanes with lower than 3/4-statute mile (1,200 m) approach visibility minimums and displaced threshold. (*U.S. Department of Transportation, Federal Aviation Administration*)

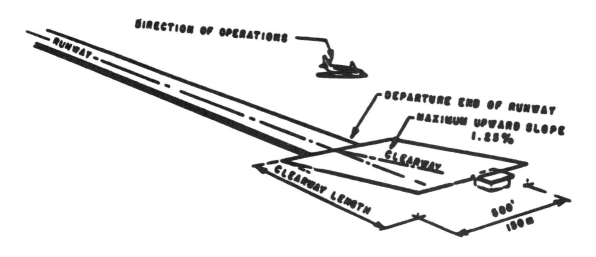

FIGURE B-49 Clearway. *(U.S. Department of Transportation, Federal Aviation Administration)*

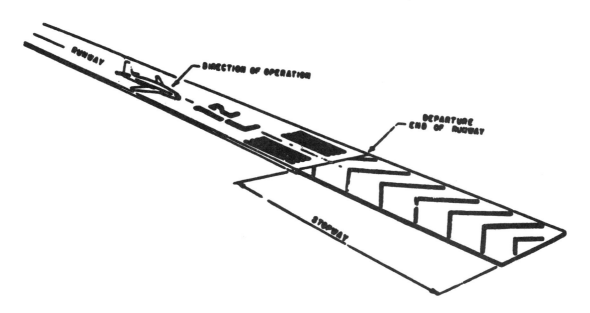

FIGURE B-49a *(continued)* Stopway. *(U.S. Department of Transportation, Federal Aviation Administration)*

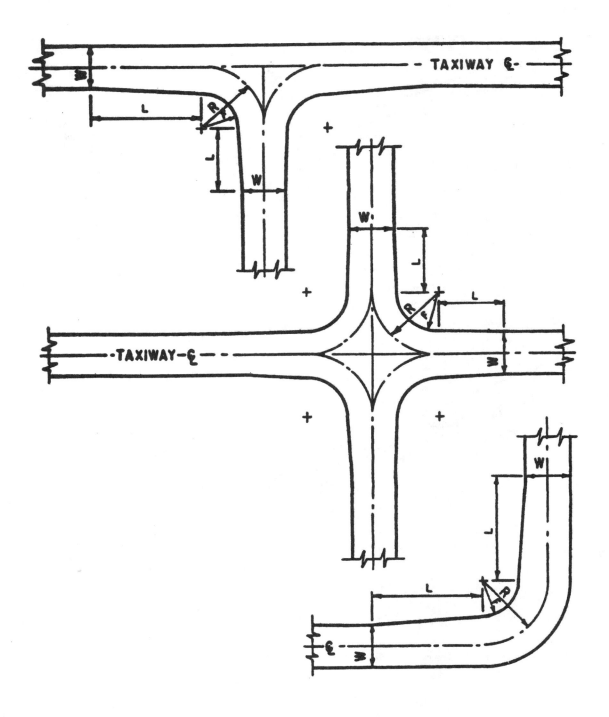

FIGURE B-50 Taxiway intersection details. *(U.S. Department of Transportation, Federal Aviation Administration)*

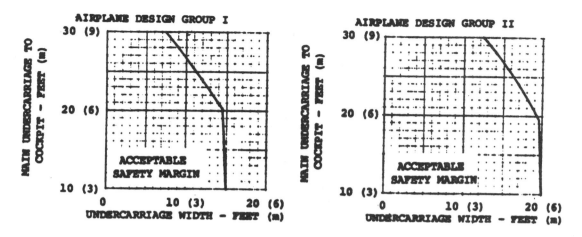

FIGURE B-51a Maintaining cockpit over center line. Airplane design groups 1 and 2. *(U.S. Department of Transportation, Fed-*

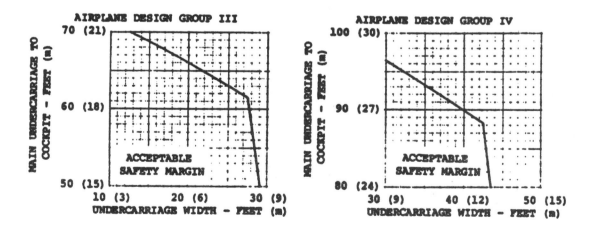

FIGURE B-51b Maintaining cockpit over center line. Airplane design groups 3 and 4. *(U.S. Department of Transportation, Federal Aviation Administration)*

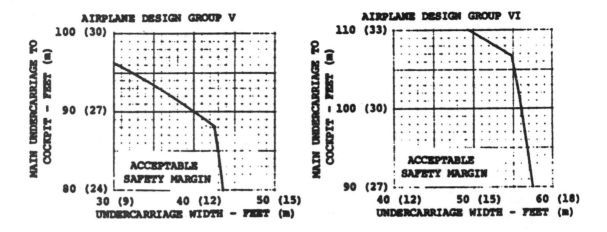

FIGURE B-51c Maintaining cockpit over center line. Airplane design groups 5 and 6. *(U.S. Department of Transportation, Federal Aviation Administration)*

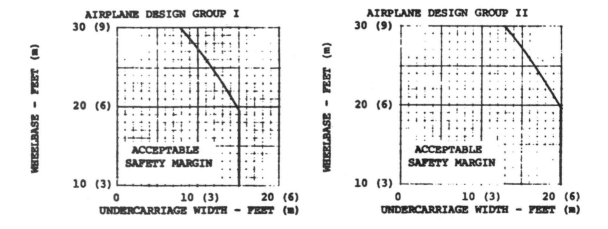

FIGURE B-52a Judgmental oversteering. Airplane design groups 1 and 2. *(U.S. Department of Transportation, Federal Avia-*

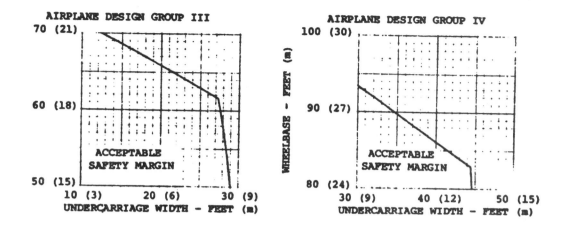

FIGURE B-52b Judgmental oversteering. Airplane design groups 3 and 4. *(U.S. Department of Transportation, Federal Aviation Administration)*

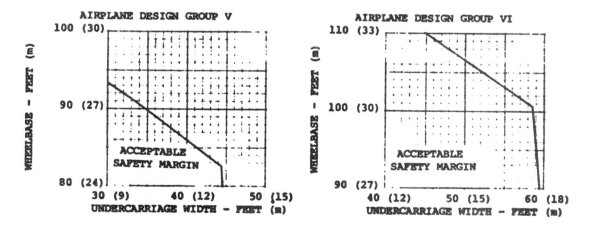

FIGURE B-52c Judgmental oversteering. Airplane design groups 5 and 6. *(U.S. Department of Transportation, Federal Aviation Administration)*

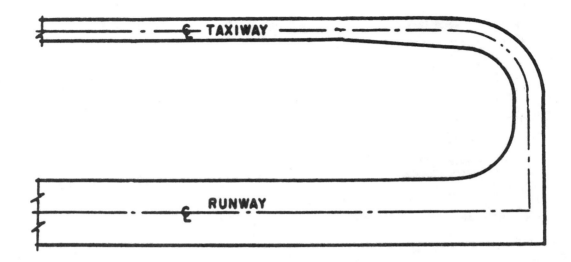

FIGURE B-53 Entrance taxiway. *(U.S. Department of Transportation, Federal Aviation Administration)*

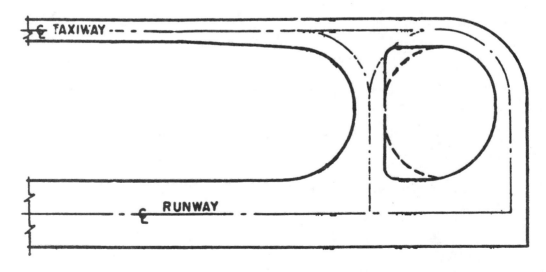

FIGURE B-54 Bypass taxiway. *(U.S. Department of Transportation, Federal Aviation Administration)*

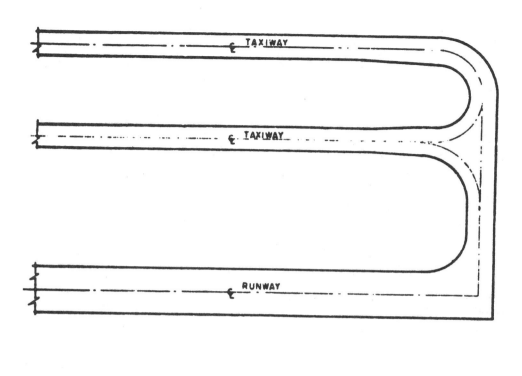

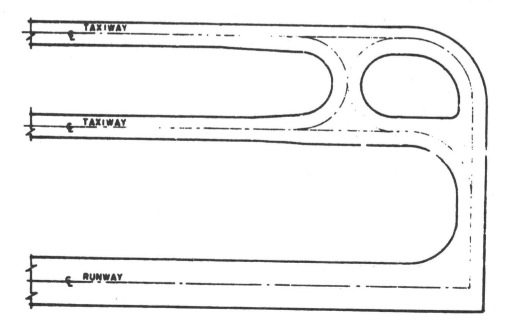

FIGURE B-55 Dual parallel taxiway entrance. *(U.S. Department of Transportation, Federal Aviation Administration)*

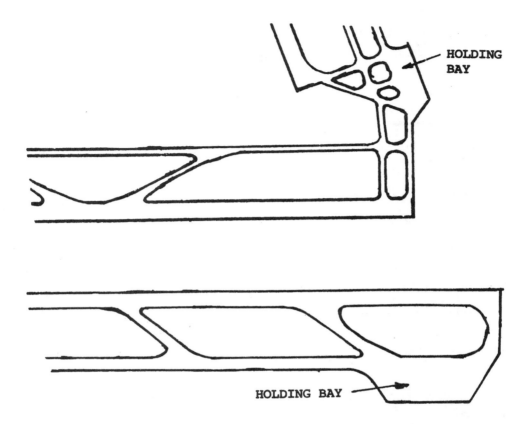

FIGURE B-56 Typical holding bay configurations. *(U.S. Department of Transportation, Federal Aviation Administration)*

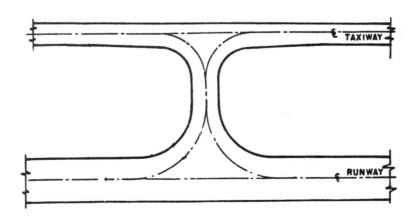

FIGURE B-57 Right-angled exit taxiway. *(U.S. Department of Transportation, Federal Aviation Administration)*

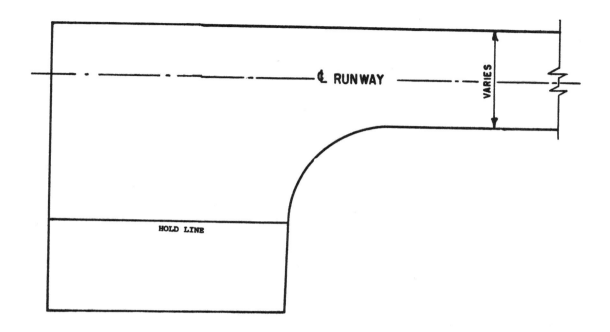

FIGURE B-58 Taxiway turnaround. *(U.S. Department of Transportation, Federal Aviation Administration)*

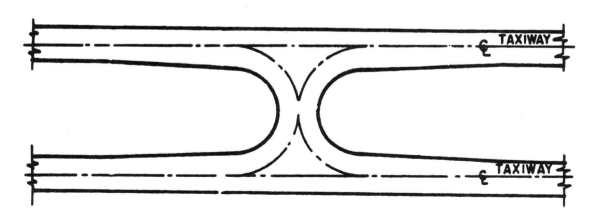

FIGURE B-59 Crossover taxiway. *(U.S. Department of Transportation, Federal Aviation Administration)*

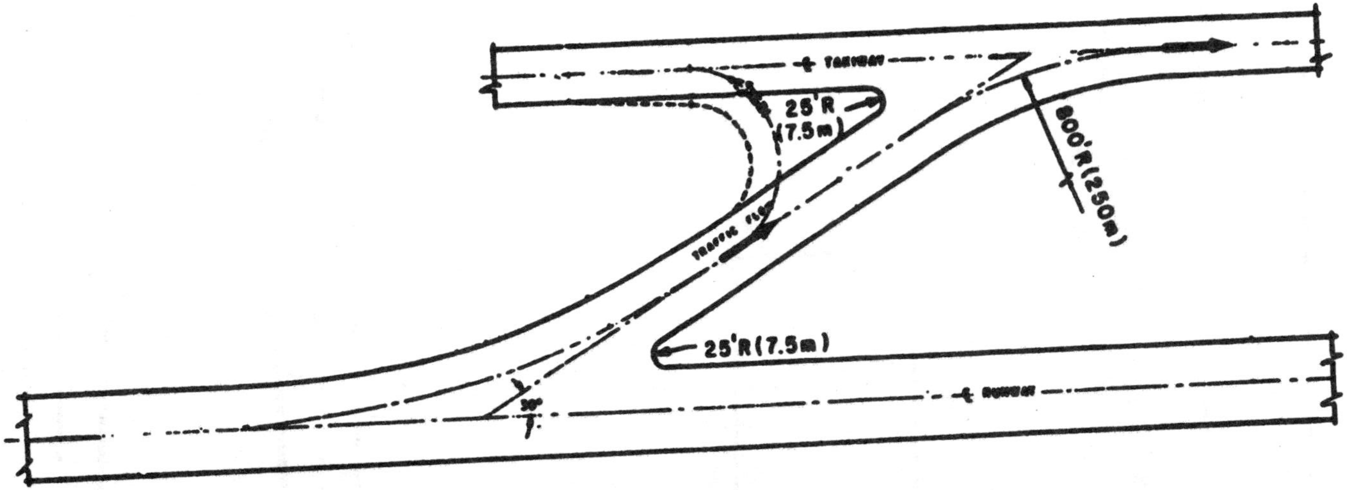

FIGURE B-60 Acute-angled exit taxiway. (*U.S. Department of Transportation, Federal Aviation Administration*)

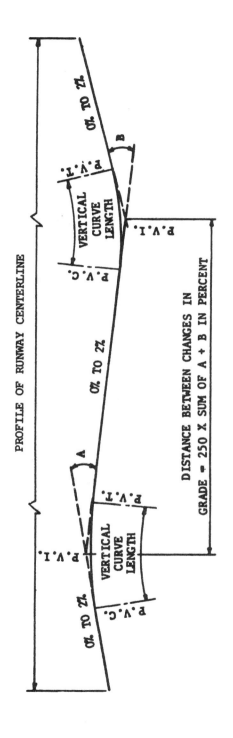

FIGURE B-61 Longitudinal grade limitations for aircraft approach categories A and B. (*U.S. Department of Transportation, Federal Aviation Administration*)

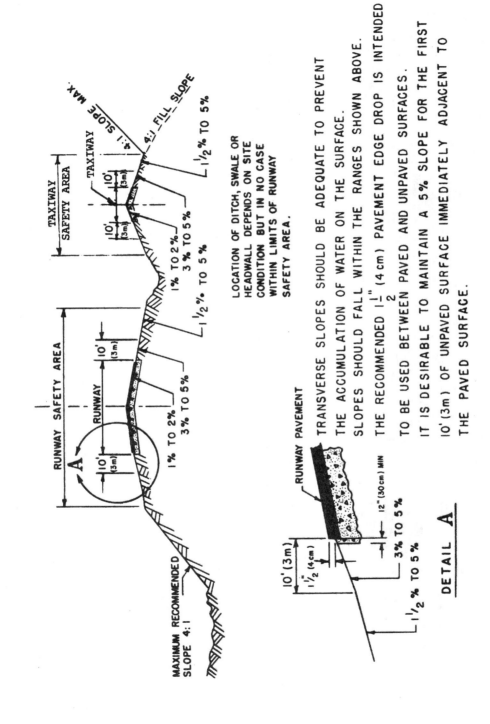

FIGURE B-62 Transverse grade limitations for aircraft approach categories A and B. (*U.S. Department of Transportation, Federal Aviation Administration*)

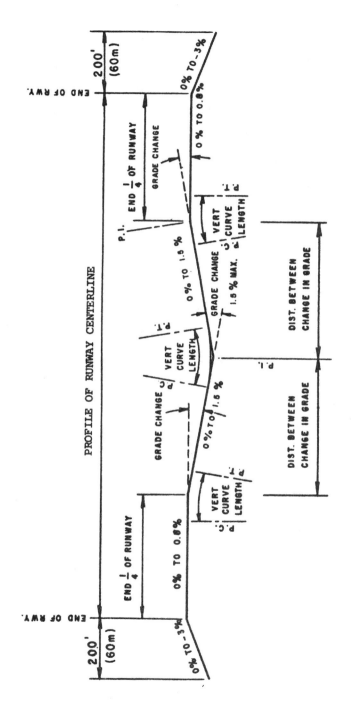

FIGURE B-63 Longitudinal grade limitations for aircraft approach categories C and D. *(U.S. Department of Transportation, Federal Aviation Administration)*

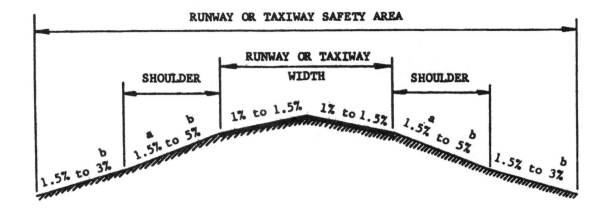

FIGURE B-64 Transverse grade limitations for aircraft approach categories C and D. *(U.S. Department of Transportation, Federal Aviation Administration)*

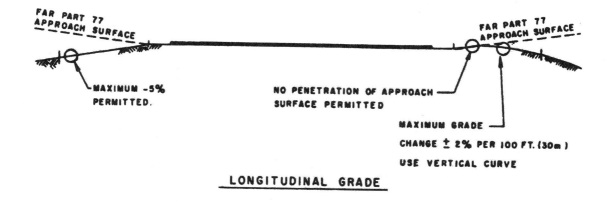

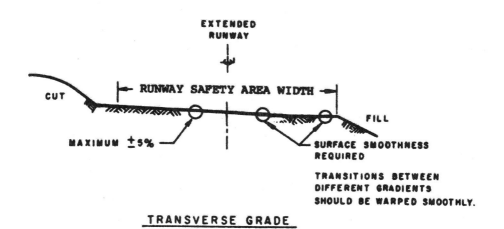

FIGURE B-65 Runway safety area grade limitations beyond 200 feet (60 m) from the runway end. *(U.S. Department of Transportation, Federal Aviation Administration)*

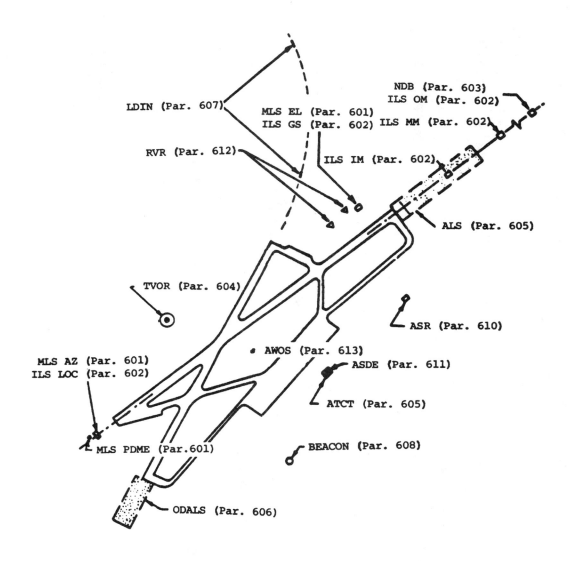

FIGURE B-66 Typical NAVAID placement. *(U.S. Department of Transportation, Federal Aviation Administration)*

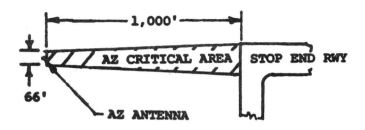

FIGURE B-67 AZ antenna critical area. *(U.S. Department of Transportation, Federal Aviation Administration)*

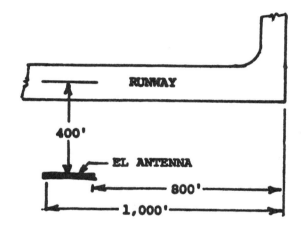

FIGURE B-68 EL antenna siting. *(U.S. Department of Transportation, Federal Aviation Administration)*

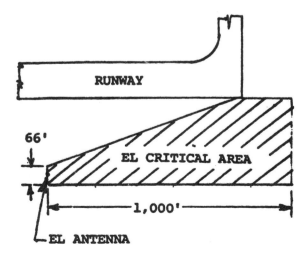

FIGURE B-69 EL antenna critical area. *(U.S. Department of Transportation, Federal Aviation Administration)*

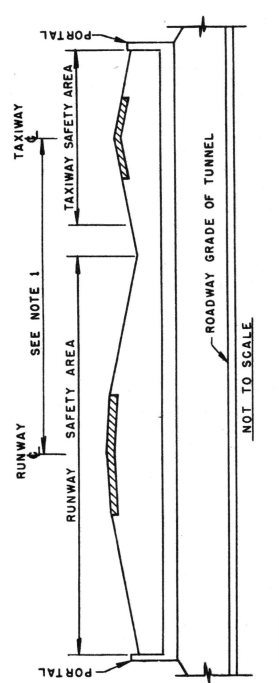

NOTES:
1. WIDTH OF TAXIWAY SAFETY AREA AND RUNWAY/TAXIWAY SEPARATION DISTANCE VARY DEPENDING ON AIRPLANE/TAXIWAY DESIGN GROUP.
2. ROADWAY TUNNEL NORMALLY HAS SLIGHT LONGITUDINAL GRADIENT AND SOME TYPE OF RETAINING WALL AT PORTALS
3. UNIFORM TUNNEL CROSS SECTION IS NORMALLY USED; AND A CONTINUOUS STRUCTURE WITHOUT OPEN SECTION IN INFIELD AREA IS PREFERRED AND RECOMMENDED WHEREVER FEASIBLE

FIGURE B-70 Cross-section full width runway-taxiway bridge. (*U.S. Department of Transportation, Federal Aviation Administration*)

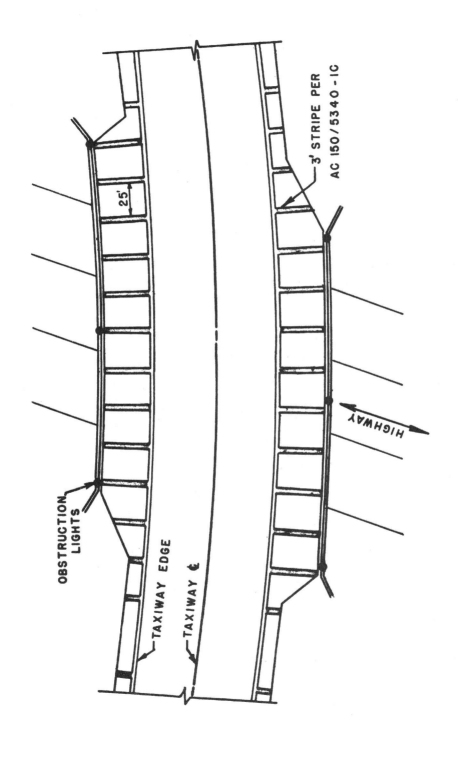

FIGURE B-71 Suggested shoulder marking of minimum width taxiway bridge. (*U.S. Department of Transportation, Federal Aviation Administration*)

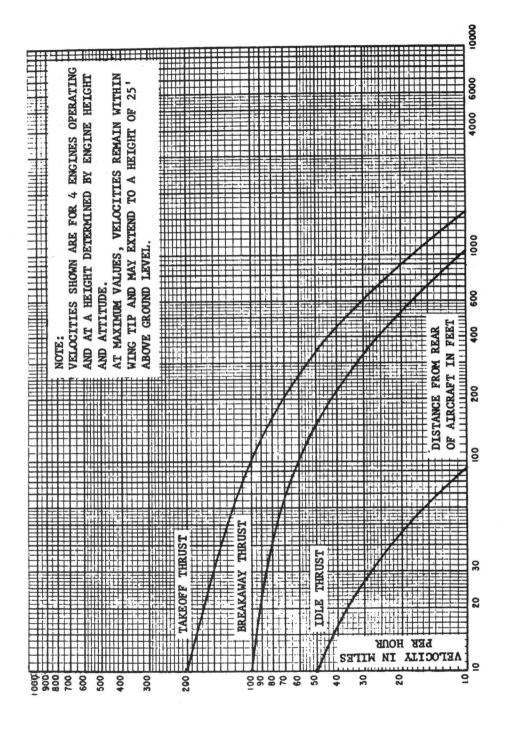

FIGURE B-72 Velocity distance curves, DC-8. *(U.S. Department of Transportation, Federal Aviation Administration)*

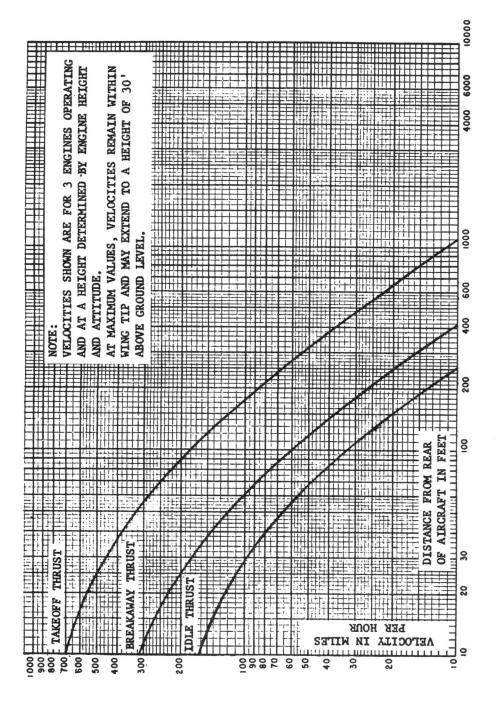

FIGURE B-73 Velocity distance curves, B-727. (*U.S. Department of Transportation, Federal Aviation Administration*)

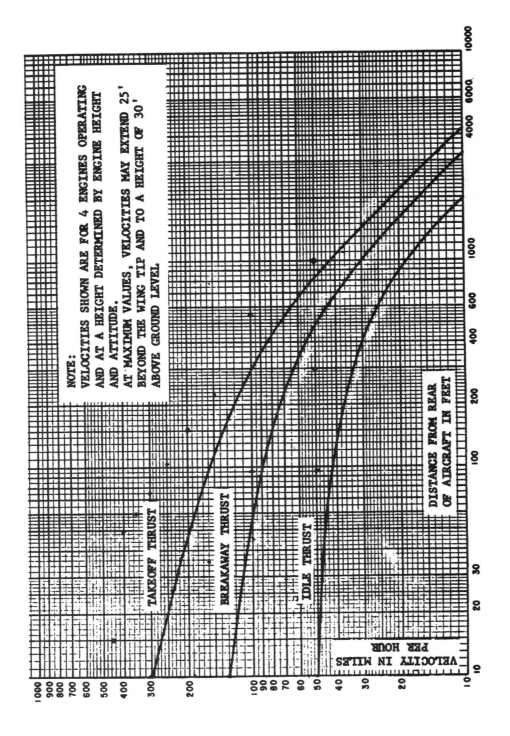

FIGURE B-74 Velocity distance curves, B-747. (*U.S. Department of Transportation, Federal Aviation Administration*)

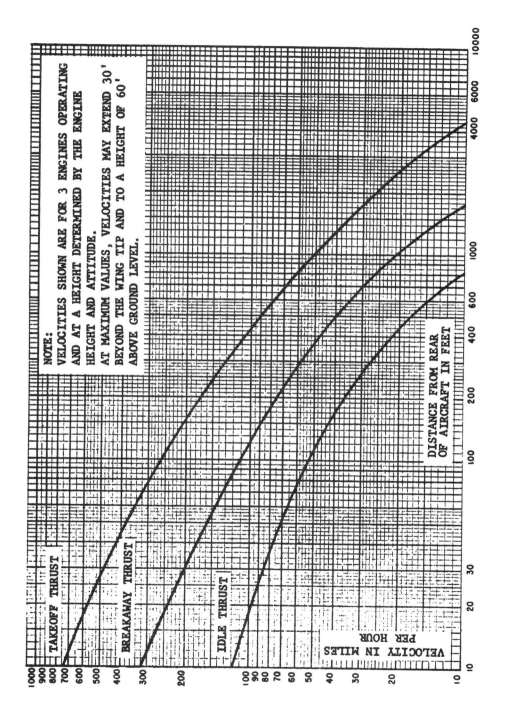

FIGURE B-75 Velocity distance curves, DC-10. (*U.S. Department of Transportation, Federal Aviation Administration*)

VELOCITY IN MILES/HOUR (KILOMETERS/HOUR)

Distance Behind Engine / Aircraft	20' (6 m)	40' (12 m)	60' (18 m)	80' (24 m)	100' (30 m)
Fan Jet Falcon					
Idle	82(132)	36(58)	25(40)	22(35)	18(29)
Breakaway 1/	150(241)	68(109)	46(74)	33(53)	27(43)
Takeoff	341(549)	155(249)	106(171)	75(121)	62(100)
Jet Commander, Lear Jet, & Hansa					
Idle	54(87)	24(39)	15(24)	11(18)	9(14)
Breakaway	114(183)	50(80)	31(50)	22(35)	18(29)
Takeoff	259(417)	114(183)	68(109)	52(84)	42(68)
Jet Star & Sabreliner					
Idle	92(148)	41(66)	25(40)	18(29)	15(24)
Breakaway	195(314)	85(137)	52(84)	39(63)	31(50)
Takeoff	443(713)	194(312)	119(192)	89(143)	72(116)
Gulfstream II					
Idle	153(246)	75(121)	48(77)	41(66)	34(55)
Breakaway	330(531)	150(241)	102(164)	72(116)	60(97)
Takeoff	750(1207)	341(549)	232(373)	164(264)	136(219)

1/ "Breakaway" is that percentage of power required to start airplanes moving and usually is approximately 55 percent of maximum continuous thrust.

FIGURE B-76 Blast velocities of business jet airplanes. *(U.S. Department of Transportation, Federal Aviation Administration)*

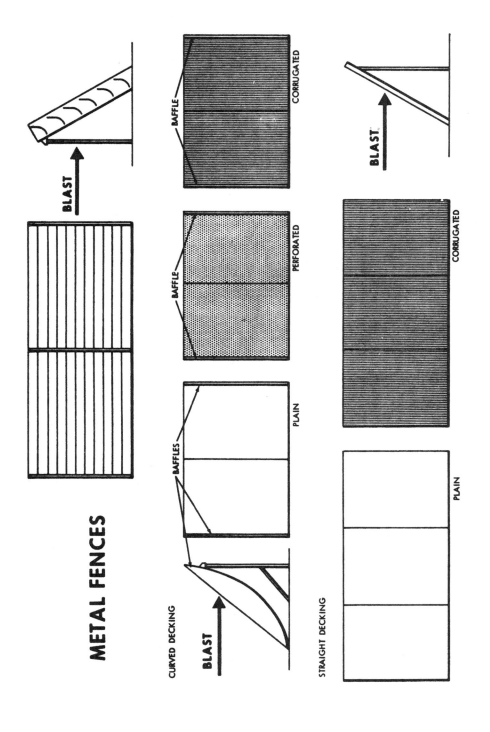

FIGURE B-77 Typical blast deflector fences, metal. *(U.S. Department of Transportation, Federal Aviation Administration)*

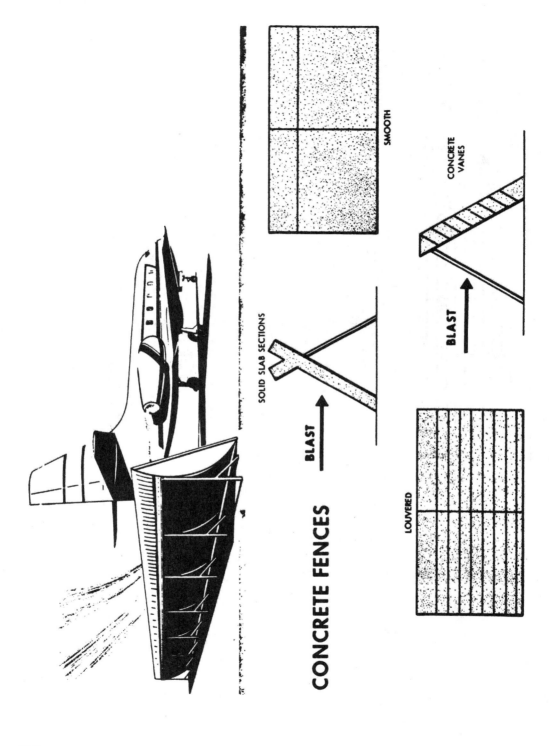

FIGURE B-78 Typical blast deflector fences, concrete. (*U.S. Department of Transportation, Federal Aviation Administration*)

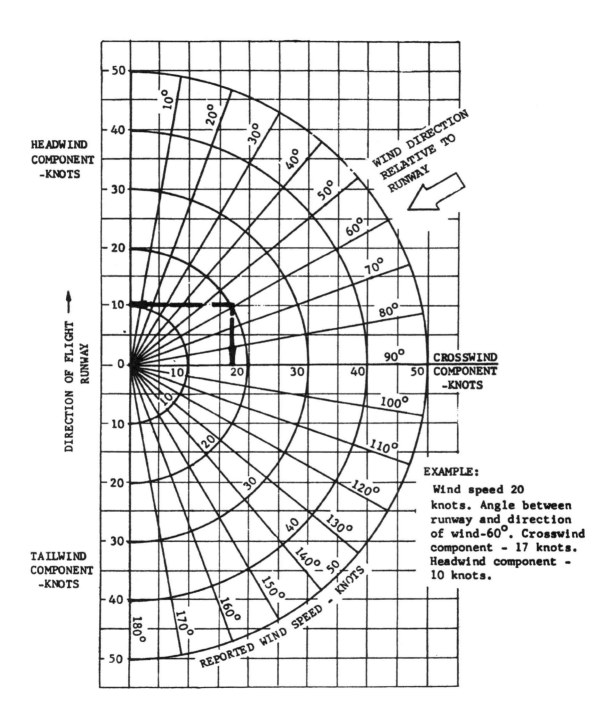

FIGURE B-79 Wind vector diagram. *(U.S. Department of Transportation, Federal Aviation Administration)*

WIND DIRECTION VERSUS WIND SPEED

STATION: Anywhere, USA   HOURS: 24 Observations/Day   PERIOD OF RECORD: 1964-1973

DIRECTION	HOURLY OBSERVATIONS OF WIND SPEED									TOTAL	AVERAGE SPEED	
	0-3	4-6	7-10	11-16	17-21	KNOTS 22-27 / MPH 25-31	28-33 / 32-38	34-40 / 39-46	41 OVER / 47 OVER		KNOTS	MPH
	0-3	4-7	8-12	13-18	19-24							
01	469	842	568	212						2091	6.2	7.1
02	568	1263	820	169						2820	6.0	6.9
03	294	775	519	73	9					1670	5.7	6.6
04	317	872	509	62	11					1771	5.7	6.6
05	268	861	437	106						1672	5.6	6.4
06	357	534	151	42	8					1092	4.9	5.6
07	369	403	273	84	36	10				1175	6.6	7.6
08	158	261	138	69	73	52	41	22		814	7.6	8.8
09	167	352	176	128	68	59	21			971	7.5	8.6
10	119	303	127	180	98	41	9			877	9.3	10.7
11	323	586	268	312	111	23	28			1651	7.9	9.1
12	618	1397	624	779	271	69	21			3779	8.3	9.6
13	472	1375	674	531	452	67				3571	8.4	9.7
14	647	1377	574	281	129					3008	6.2	7.1
15	338	1093	348	135	27					1941	5.6	6.4
16	560	1399	523	121	19					2622	5.5	6.3
17	587	883	469	128	12					2079	5.4	6.2
18	1046	1984	1068	297	83	18				4496	5.8	6.7
19	499	793	586	241	92					2211	6.2	7.1
20	371	946	615	243	64					2239	6.6	7.6
21	340	732	528	323	147	8				2078	7.6	8.8
22	479	768	603	231	115	38	19			2253	7.7	8.9
23	187	1008	915	413	192					2715	7.9	9.1
24	458	943	800	453	96	11	18			2779	7.2	8.2
25	351	899	752	297	102	21	9			2431	7.2	8.2
26	368	731	379	208	53					1739	6.3	7.2
27	411	748	469	232	118	19				1997	6.7	7.7
28	191	554	276	287	118					1426	7.3	8.4
29	271	642	548	479	143	17				2100	8.0	9.3
30	379	873	526	543	208	34				2563	8.0	9.3
31	299	643	597	618	222	19				2398	8.5	9.8
32	397	852	521	559	158	23				2510	7.9	9.1
33	236	721	324	238	48					1567	6.7	7.7
34	280	916	845	307	24					2372	6.9	7.9
35	252	931	918	487	23					2611	6.9	7.9
36	501	1568	1381	569	27					4046	7.0	8.0
00	7729									7729	0.0	0.0
TOTAL	21676	31828	19849	10437	3357	529	166	22		87864	6.9	7.9

FIGURE B-80 Typical environment data service wind summary. *(U.S. Department of Transportation, Federal Aviation Administration)*

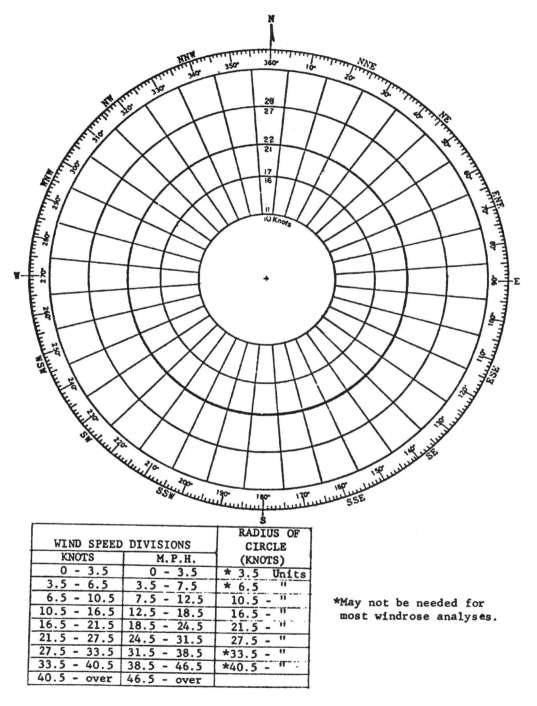

FIGURE B-81 Windrose blank showing direction and divisions. *(U.S. Department of Transportation, Federal Aviation Administration)*

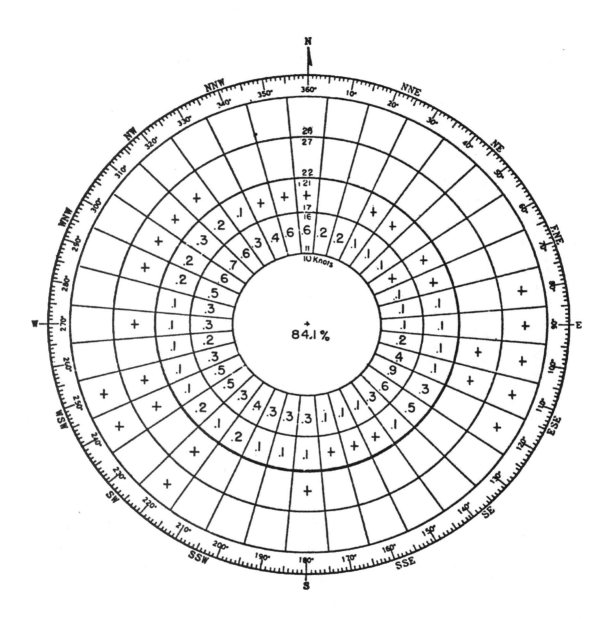

FIGURE B-82 Completed windrose using figure B-80 data. *(U.S. Department of Transportation, Federal Aviation Administration)*

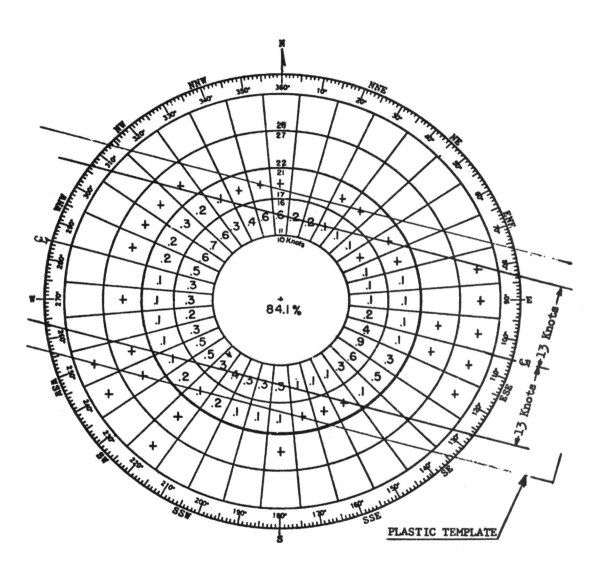

FIGURE B-83 Windrose analysis. *(U.S. Department of Transportation, Federal Aviation Administration)*

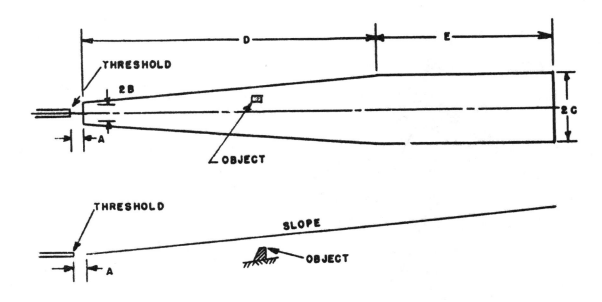
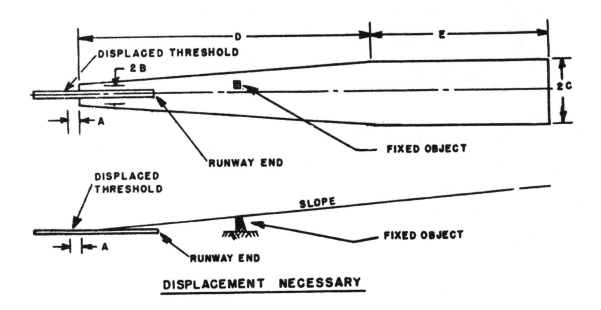

FIGURE B-84 Approach slopes. *(U.S. Department of Transportation, Federal Aviation Administration)*

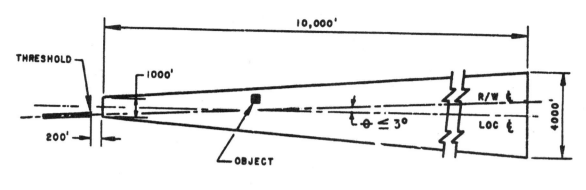
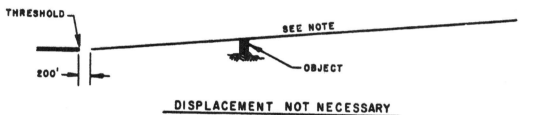

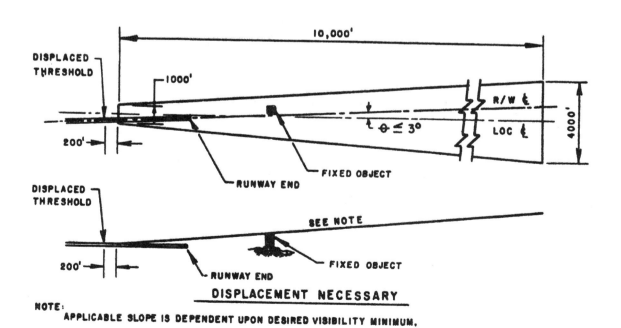

FIGURE B-85 Approach slopes—offset localizer. *(U.S. Department of Transportation, Federal Aviation Administration)*

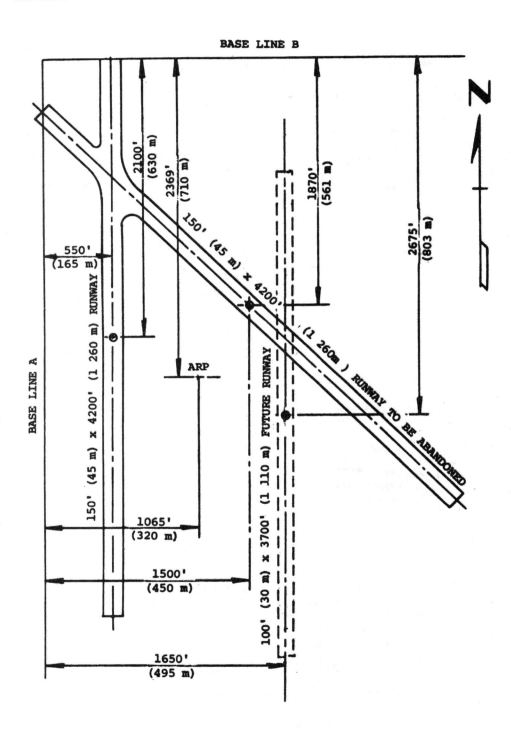

FIGURE B-86 Sample layout.

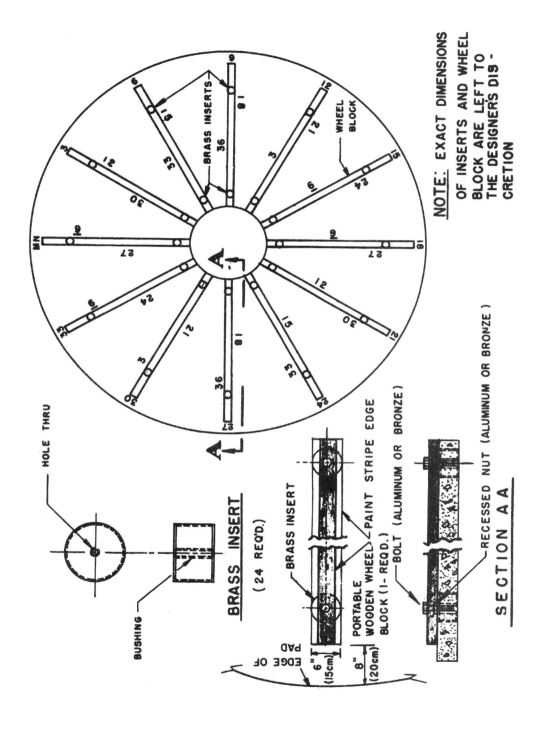

FIGURE B-87 Marking layout and details of wheel block.

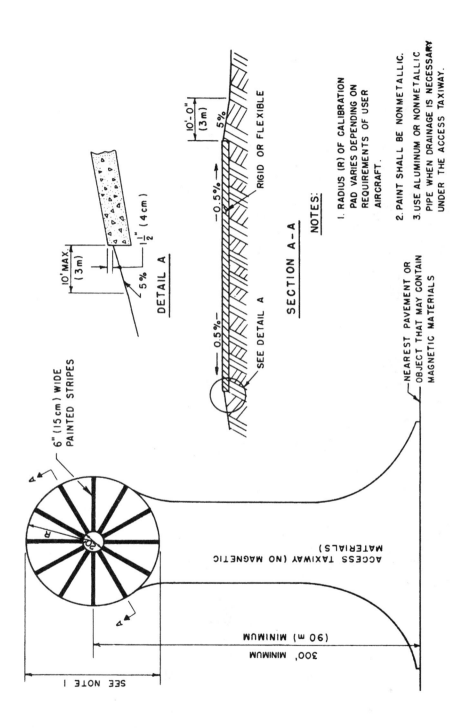

FIGURE B-88 Type 1. compass calibration pad.

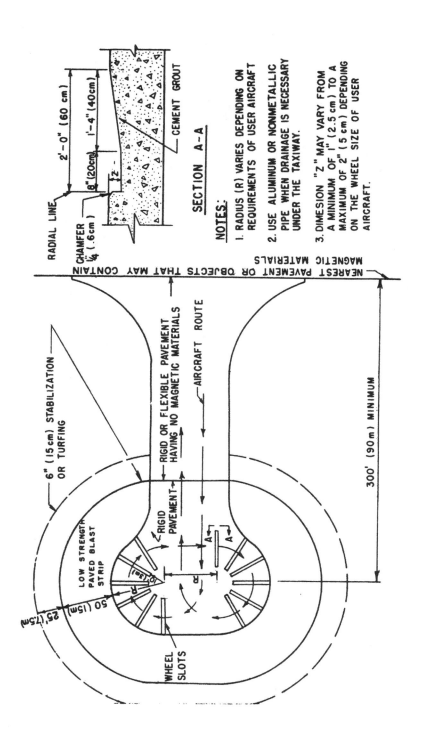

FIGURE B-89 Type 2. compass calibration pad.

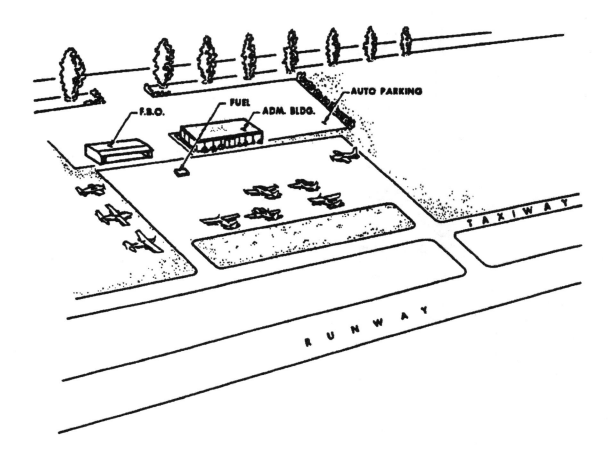

FIGURE B-90 Parking apron area.

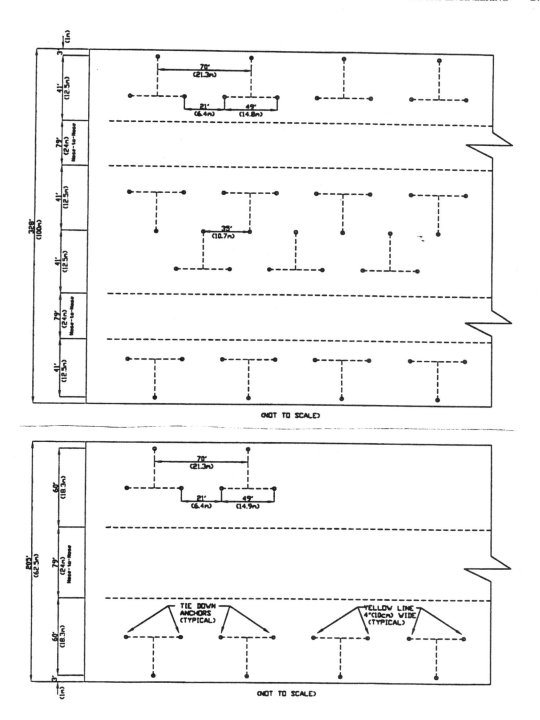

FIGURE B-91 Tiedown layouts.

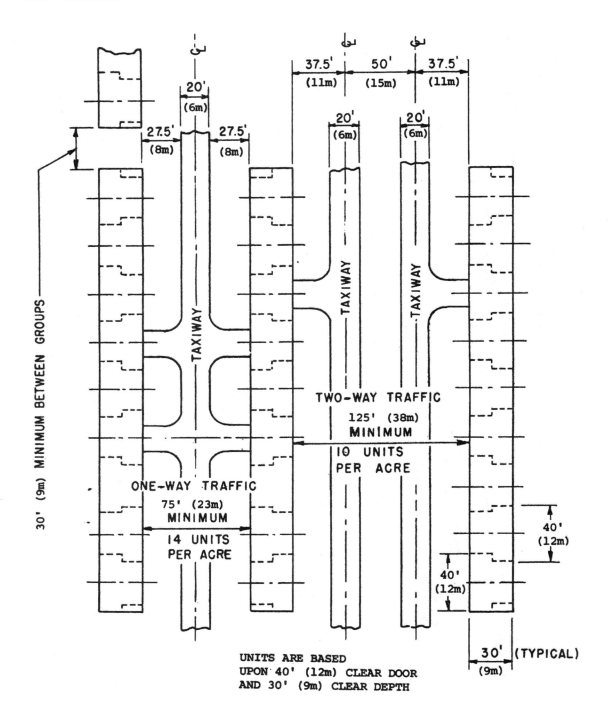

FIGURE B-92 T-hanger layout.

Exit taxiway cumulative utilization percentages

DISTANCE THRESHOLD TO EXIT	WET RUNWAYS RIGHT & ACUTE ANGLED EXITS				DRY RUNWAYS RIGHT ANGLED EXITS				ACUTE ANGLED EXITS			
	A	B	C	D	A	B	C	D	A	B	C	D
0 ft (0 m)	0	0	0	0	0	0	0	0	0	0	0	0
500 ft (152)	0	0	0	0	0	0	0	0	1	0	0	0
1000 ft (305 m)	4	0	0	0	6	0	0	0	13	0	0	0
1500 ft (457 m)	23	0	0	0	39	0	0	0	53	0	0	0
2000 ft (610 m)	60	0	0	0	84	1	0	0	90	1	0	0
2500 ft (762 m)	84	1	0	0	99	10	0	0	99	10	0	0
3000 ft (914 m)	96	10	0	0	100	39	0	0	100	40	0	0
3500 ft (1067 m)	99	41	0	0	100	81	2	0	100	82	9	0
4000 ft (1219 m)	100	80	1	0	100	98	8	0	100	98	26	3
4500 ft (1372 m)	100	97	4	0	100	100	24	2	100	100	51	19
5000 ft (1524 m)	100	100	12	0	100	100	49	9	100	100	76	55
5500 ft (1676 m)	100	100	27	0	100	100	75	24	100	100	92	81
6000 ft (1829 m)	100	100	48	10	100	100	92	71	100	100	98	95
6500 ft (1981 m)	100	100	71	35	100	100	98	90	100	100	100	99
7000 ft (2134 m)	100	100	88	64	100	100	100	98	100	100	100	100
7500 ft (2686 m)	100	100	97	84	100	100	100	100	100	100	100	100
8000 ft (2438 m)	100	100	100	93	100	100	100	100	100	100	100	100
8500 ft (2591 m)	100	100	100	99	100	100	100	100	100	100	100	100
9000 ft (2743 m)	100	100	100	100	100	100	100	100	100	100	100	100

A - Small, single engine 12,500 lbs (5 700 kg) or less
B - Small, twin engine 12,500 lbs (5 700 kg) or less
C - Large 12,500 lbs (5 700 kg) to 300,000 lbs (136 000 kg)
D - Heavy 300,000 lbs (136 000 kg)

FIGURE B-93 Exit taxiway cumulative utilization percentages.

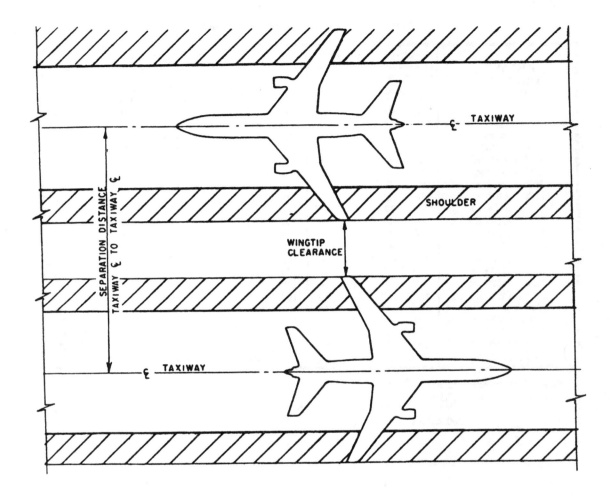

FIGURE B-94 Wingtip clearance—parallel taxiways.

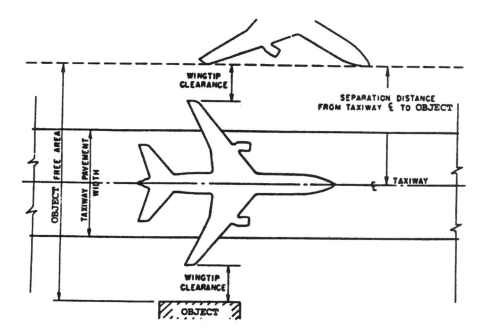

FIGURE B-95 Wingtip clearance from taxiway.

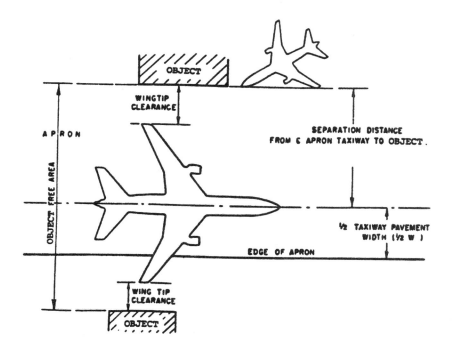

FIGURE B-96 Wingtip clearance from apron taxiway.

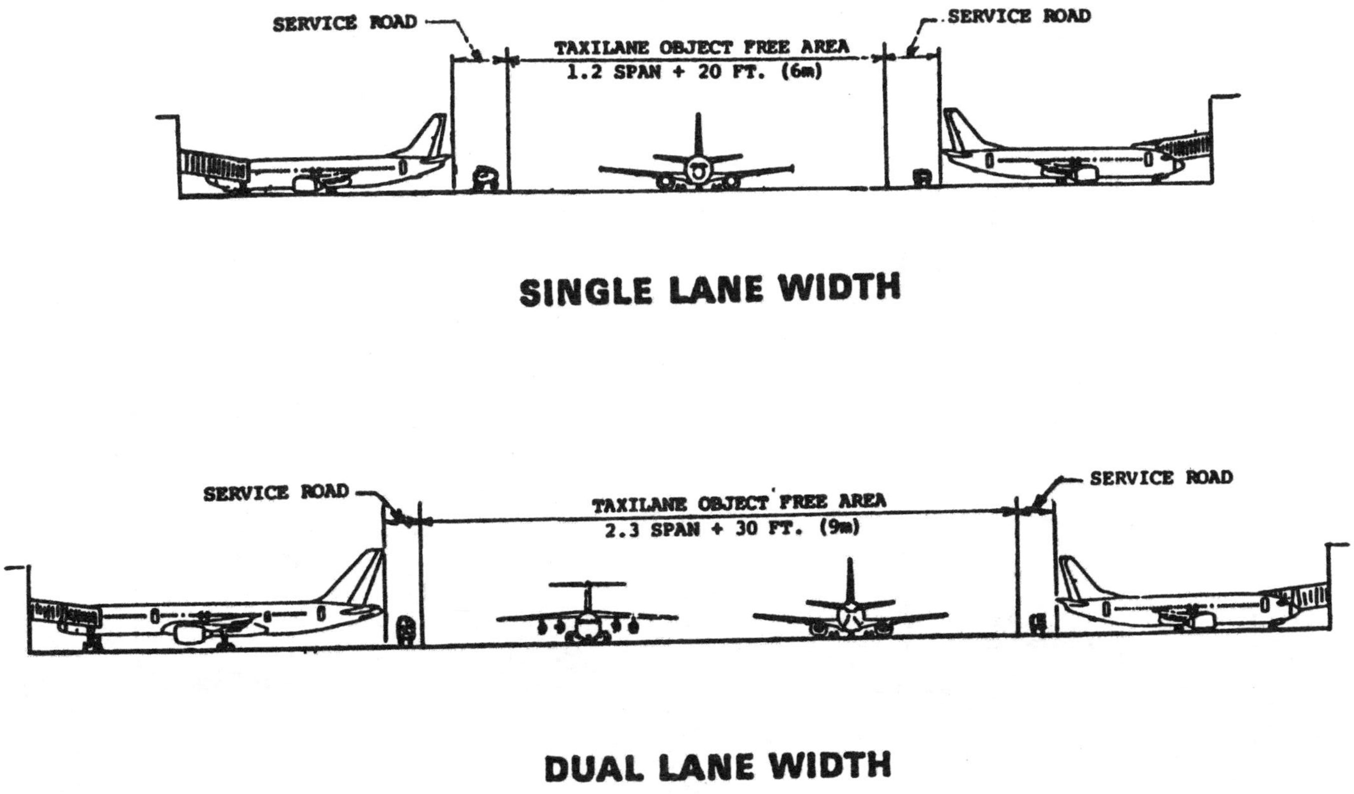
SERVICE ROAD
TAXILANE OBJECT FREE AREA
1.2 SPAN + 20 FT. (6m)
SERVICE ROAD
SINGLE LANE WIDTH
SERVICE ROAD
TAXILANE OBJECT FREE AREA
2.3 SPAN + 30 FT. (9m)
SERVICE ROAD
DUAL LANE WIDTH

FIGURE B-97 Wingtip clearance from taxilane.

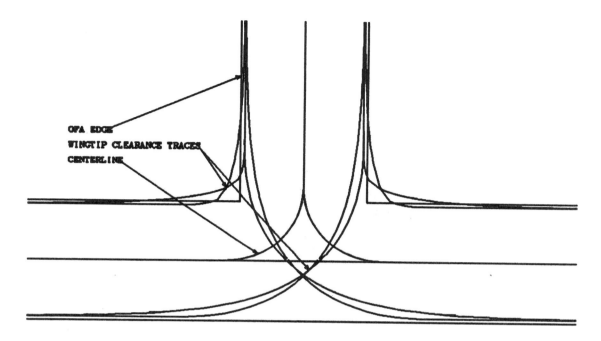

FIGURE B-98 McDonnell-Douglas MD-88 wingtip clearance trace for a 100-foot (30.5 m) radius centerline.

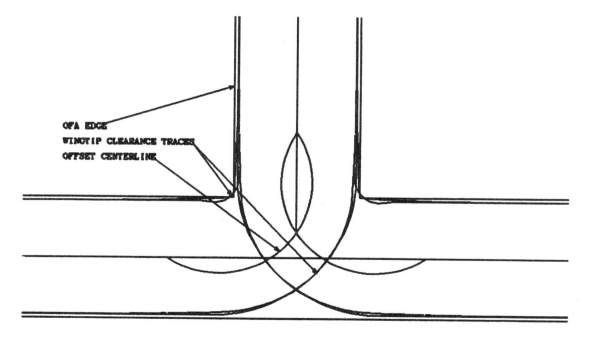

FIGURE B-99 McDonnell-Douglas MD-88 wingtip clearance trace for a 120-foot (36.5 m) radius centerline.

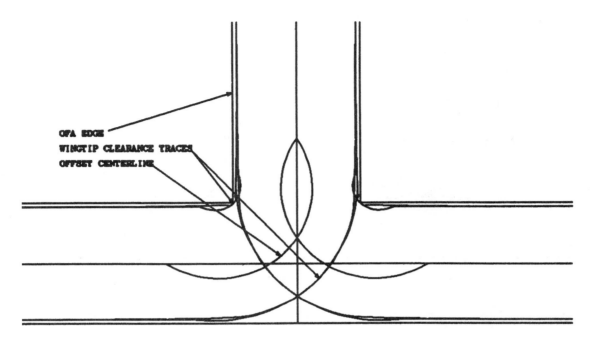

FIGURE B-100 Boeing 727-200 wingtip clearance trace for a 120-foot (36.5) radius offset centerline.

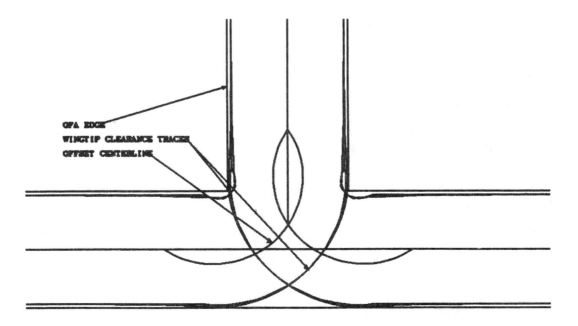

FIGURE B-101 Boeing 727-100 wingtip clearance trace for a 120-foot (36.5) radius offset centerline.

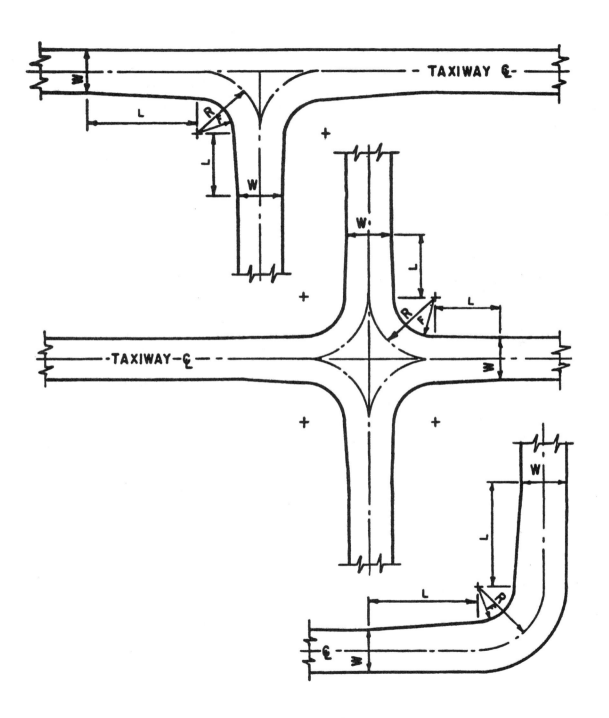

FIGURE B-102 Taxiway intersection details.

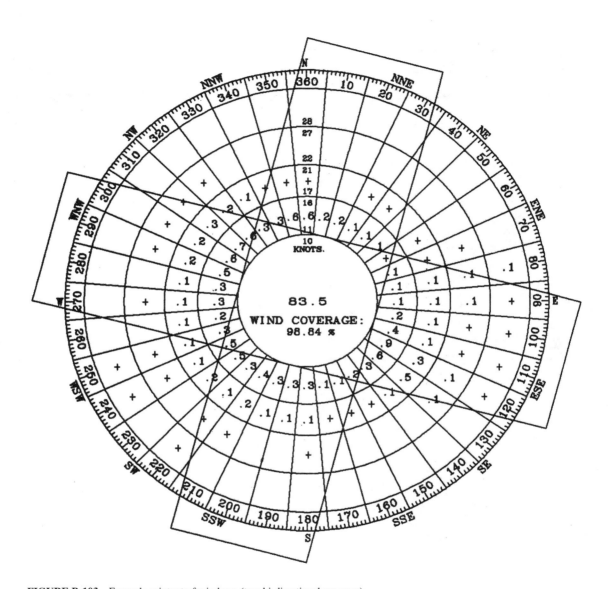

FIGURE B-103 Example printout of windrose (two bi-directional runways).

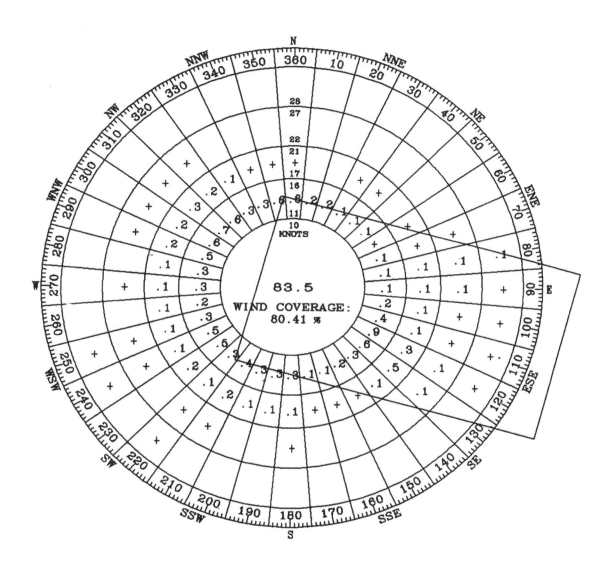

FIGURE B-104 Example printout of windrose (one uni-directional runway).

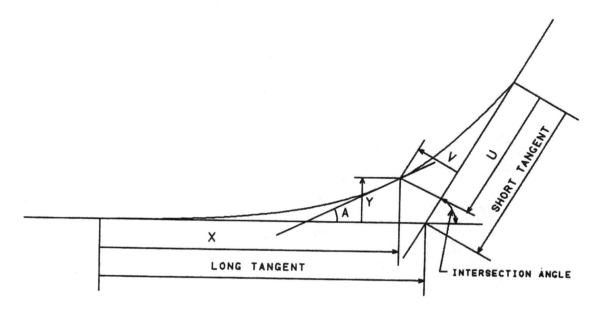

FIGURE B-105 Nomenclature used in the taxiway design task.

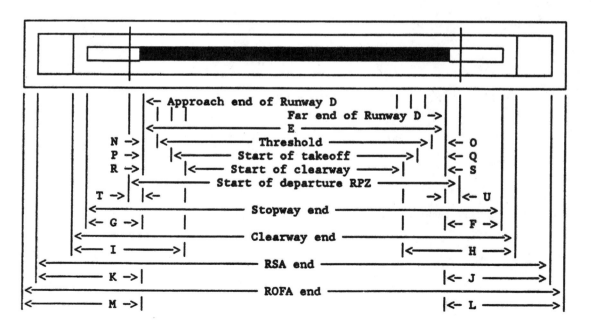

FIGURE B-106 Nomenclature used in the declared distance task.

AIRPLANES ARRANGED BY AIRPLANE MANUFACTURER, AND AIRPORT REFERENCE CODE

Section 1. Alphabetical Listing (U.S. customary units)

Aircraft	Airport Reference Code	Appch Speed Knots	Wingspan Feet	Length Feet	Tail Height Feet	Maximum Takeoff Lbs
Aeritalia G-222	B-III	109	93.8	74.4	32.0	61,700
Aerocom Skyliner	A-II	88	54.0	54.3	16.5	12,500
Aerospatiale C 160 Trans.	C-IV	124	131.3	106.3	38.7	108,596
Aerospatiale NORD-262	B-II	96	71.9	63.3	20.4	23,480
Aerospatiale SE 210 Carav.	C-III	127	112.5	105.0	28.6	114,640
Aerospatiale SN 601 Corv.	B-I	118	42.2	45.4	13.9	14,550
Ahrens AR 404	B-II	98	66.0	52.7	19.0	18,500
AIDC/CAF XC-2	A-III	86	81.7	65.9	25.3	27,500
Airbus A-300-600	C-IV	135	147.1	177.5	54.7	363,763
Airbus A-300-B4	C-IV	132	147.1	175.5	55.5	330,700
Airbus A-310-300	C-IV	125	144.1	153.2	52.3	330,693
Airbus A-320-100	C-III	138	111.3	123.3	39.1	145,505
Air-Metal AM-C 111	B-II	96	63.0	55.2	21.0	18,629
AJI Hustler 400	B-I	98	28.0	34.8	9.8	6,000
Antonov AN-10	C-IV	126	124.8	121.4	32.2	121,500
Antonov AN-12	C-IV	127	124.8	109.0	34.6	121,500
Antonov AN-124	C-VI	124	232.0	223.0	66.2	800,000
Antonov AN-14	A-II	52	72.1	37.2	15.2	7,607
Antonov AN-22	C-V	140 *	211.0	167.0	41.2	500,000
Antonov AN-24	B-III	119	95.8	77.2	27.3	46,305
Antonov AN-26	C-III	121	95.8	78.1	28.1	52,920
Antonov AN-28	A-II	88	72.1	42.6	16.1	12,350
Antonov AN-30	B-III	112	96.4	80.1	27.3	51,040
Antonov AN-72	A-III	89 *	84.7	84.7	27.0	66,000
AW.650 Argosy 220	C-III	123	115.0	86.8	27.0	93,000
AW.660 Argosy C.Mk.1	B-III	113	115.0	89.1	27.0	97,000
BAC 111-200	C-III	129	88.5	93.5	24.5	79,000
BAC 111-300	C-III	128	88.5	93.5	24.5	88,500
BAC 111-400	C-III	137	88.5	93.5	24.5	87,000
BAC 111-475	C-III	135	93.5	93.5	24.5	98,500
BAC 111-500	D-III	144	93.5	107.0	24.5	104,500
BAC/Aerospatiale Concord	D-III	162	83.8	205.4	37.4	408,000
BAe 146-100	B-III	113	86.4	85.8	28.3	74,600
BAe 146-200	B-III	117	86.4	93.7	28.3	88,250
BAe 146-300	C-III	121	86.4	104.2	28.1	104,000
BAe Jetstream 31	B-II	99	52.0	47.2	17.5	14,550
Beech Airliner 1900-C	B-II	120 *	54.5	57.8	14.9	16,600
Beech Airliner C99	B-I	107	45.9	44.6	14.4	11,300
Beech Baron 58	B-I	96	37.8	29.8	9.8	5,500
Beech Baron 58P	B-I	101	37.8	29.8	9.1	6,200
Beech Baron 58TC	B-I	101	37.8	29.8	9.1	6,200
Beech Baron B55	A-I	90	37.8	28.0	9.1	5,100
Beech Baron E55	A-I	88	37.8	29.0	9.1	5,300
Beech Bonanza A36	A-I	72	33.5	27.5	8.6	3,650

FIGURE B-107 Airplanes arranged by airplane manufacturer, and airport reference code.

Aircraft	Airport Reference Code	Appch Speed Knots	Wingspan Feet	Length Feet	Tail Height Feet	Maximum Takeoff Lbs
Beech Bonanza B36TC	A-I	75	37.8	27.5	8.6	3,850
Beech Bonanza F33A	A-I	70	33.5	26.7	8.2	3,400
Beech Bonanza V35B	A-I	70	33.5	26.4	6.6	3,400
Beech Duchess 76	A-I	76	38.0	29.0	9.5	3,900
Beech Duke B60	B-I	98	39.2	33.8	12.3	6,775
Beech E18S	A-II	87	49.7	35.2	9.5	9,300
Beech King Air B100	B-I	111	45.8	39.9	15.3	11,800
Beech King Air C90-1	B-II	100	50.2	35.5	14.2	9,650
Beech King Air F90	B-I	108	45.9	39.8	15.1	10,950
Beech Sierra 200-B24R	A-I	70	32.8	25.7	8.2	2,750
Beech Skipper 77	A-I	63	30.0	24.0	6.9	1,675
Beech Sundowner 180-C23	A-I	68	32.8	25.7	8.2	2,450
Beech Super King Air B200	B-II	103	54.5	43.8	15.0	12,500
BN-2A Mk.3 Trislander	A-II	65	53.0	45.7	14.2	10,000
Boeing 707-100	C-IV	139	130.8	145.1	41.7	257,340
Boeing 707-200	D-IV	145	130.8	145.1	41.7	257,340
Boeing 707-320	C-IV	139	142.4	152.9	42.2	312,000
Boeing 707-320B	C-IV	136	145.8	152.9	42.1	336,600
Boeing 707-420	C-IV	132	142.4	152.9	42.2	312,000
Boeing 720	C-IV	133	130.8	136.2	41.4	229,300
Boeing 720B	C-IV	137	130.8	136.8	41.2	234,300
Boeing 727-100	C-III	125	108.0	133.2	34.3	169,000
Boeing 727-200	C-III	138	108.0	153.2	34.9	209,500
Boeing 737-100	C-III	137	93.0	94.0	37.2	110,000
Boeing 737-200	C-III	137	93.0	100.2	37.3	115,500
Boeing 737-300	C-III	137	94.8	109.6	36.6	135,000
Boeing 737-400	C-III	139	94.8	119.6	36.6	150,000
Boeing 737-500	C-III	140 *	94.8	101.8	36.6	133,500
Boeing 747-100	D-V	152	195.7	231.8	64.3	600,000
Boeing 747-200	D-V	152	195.7	231.8	64.7	833,000
Boeing 747-300SR	D-V	141	195.7	231.8	64.3	600,000
Boeing 747-400	D-V	154	213.0	231.8	64.3	870,000
Boeing 747-SP	C-V	140	195.7	184.8	65.8	696,000
Boeing 757	C-IV	135	124.8	155.3	45.1	255,000
Boeing 767-200	C-IV	130	156.1	159.2	52.9	315,000
Boeing 767-300	C-IV	130	156.1	180.3	52.6	350,000
Boeing 777	D-IV	145	155.0	181.5	44.8	380,000
Boeing B-52	D-V	141 *	185.0	157.6	40.8	488,000
Boeing C97 Stratocruiser	B-IV	105	141.3	110.3	38.3	145,800
Boeing E-3	C-IV	137	145.9	153.0	42.0	325,000
Boeing E-4 (747-200)	D-V	152	195.7	231.8	64.7	833,000
Boeing YC-14	A-IV	89	129.0	131.7	48.3	216,000
Bristol Brittania 300/310	B-IV	117	142.3	124.2	37.5	185,000
Canadair CL-44	C-IV	123	142.3	136.8	38.4	210,000
Canadair CL-600	C-II	125	61.8	68.4	20.7	41,250
Casa C-207A Azor	B-III	102	91.2	68.4	25.4	36,400
Casa C-212-200 Aviocar	A-II	81	62.3	49.8	20.7	16,976
Cessna Citation I	B-I	108	47.1	43.5	14.3	11,850
Cessna Citation II	B-II	108	51.7	47.2	15.0	13,300
Cessna Citation III	B-II	114	53.5	55.5	16.8	22,000
Cessna-150	A-I	55	32.7	23.8	8.0	1,600

FIGURE B-107 *(continued)* Airplanes arranged by airplane manufacturer, and airport reference code.

Aircraft	Airport Reference Code	Appch Speed Knots	Wingspan Feet	Length Feet	Tail Height Feet	Maximum Takeoff Lbs
Cessna-177 Cardinal	A-I	64	35.5	27.2	8.5	2,500
Cessna-402 Businessliner	B-I	95	39.8	36.1	11.6	6,300
Cessna-404 Titan	B-I	92	46.3	39.5	13.2	8,400
Cessna-414 Chancellor	B-I	94	44.1	36.4	11.5	6,785
Cessna-421 Golden Eagle	B-I	96	41.7	36.1	11.6	7,450
Cessna-441 Conquest	B-II	100	49.3	39.0	13.1	9,925
Convair 240	B-III	107	91.8	74.7	26.9	41,790
Convair 340	B-III	104	105.3	81.5	28.2	49,100
Convair 440	B-III	106	105.3	81.5	28.2	49,100
Convair 580	B-III	107	105.3	81.5	29.2	54,600
Dassault 1150 Atlantic	C-IV	130 *	122.7	104.2	37.2	100,000
Dassault 941	A-II	59	76.7	77.9	30.7	58,400
Dassault FAL-10	B-I	104	42.9	45.5	15.1	18,740
Dassault FAL-20	B-II	107	53.5	56.3	17.4	28,660
Dassault FAL-200	B-II	114	53.5	56.3	17.4	30,650
Dassault FAL-50	B-II	113	61.9	60.8	22.9	37,480
Dassault FAL-900	B-II	100	63.4	66.3	24.8	45,500
Dassault Mercure	B-III	117	100.2	114.3	37.3	124,500
DHC-2 Beaver	A-I	50	48.0	30.3	9.0	5,100
DHC-4 Caribou	A-III	77	95.6	72.6	31.8	28,500
DHC-5D Buffalo	B-III	91	96.0	79.0	28.7	49,200
DHC-6-300 Twin Otter	A-II	75	65.0	51.7	19.5	12,500
DHC-7 Dash 7-100	A-III	83	93.0	80.7	26.2	43,000
DHC-8 Dash 8-300	A-III	90	90.0	84.3	24.6	41,100
DH.104 Dove 8	A-II	84	57.0	39.2	13.3	8,950
DH.106 Comet 4C	B-III	108	115.0	118.0	29.5	162,000
DH.114 Heron 2	A-II	85	71.5	48.5	15.6	13,500
Dornier DO 28D-2	A-II	74	51.0	37.4	12.8	8,855
Dornier LTA	A-II	74 *	58.4	54.4	18.2	15,100
Embraer-110 Bandeirante	B-II	92	50.3	49.5	16.5	13,007
Embraer-121 Xingu	B-I	92	47.4	40.2	15.9	12,500
Embraer-326 Xavante	B-I	102	35.6	34.9	12.2	11,500
Embraer-820 Navajo Chief	A-I	74	40.7	34.6	13.0	7,000
Fairchild C-119	C-III	122	109.3	86.5	27.5	77,000
Fairchild C-121	A-III	88	110.0	75.8	34.1	60,000
Fairchild FH-227 B,D	B-III	105	95.2	83.1	27.5	45,500
Fairchild F-27 A,J	B-III	109	95.2	77.2	27.5	42,000
FMA IA-50 Guarni II	B-II	101	64.1	48.8	19.1	15,700
Fokker F-27-500	B-III	102	95.2	82.3	29.3	45,000
Fokker F-28-1000	B-II	119	77.3	89.9	27.8	65,000
Fokker F-28-2000	B-II	119	77.3	97.2	27.8	65,000
Fokker F-28-3000	C-III	121	82.3	89.9	27.8	73,000
Fokker F-28-4000	C-III	121	82.3	97.2	27.8	73,000
Fokker F-28-6000	B-III	113	82.3	97.2	27.8	73,000
Foxjet ST-600-8	B-I	97	31.6	31.8	10.2	4,550
GAC-100	A-II	86	70.0	67.3	24.9	28,900
Gates Learjet 24	C-I	128	35.6	43.3	12.6	13,000
Gates Learjet 25	C-I	137	35.6	47.6	12.6	15,000
Gates Learjet 28/29	B-I	120	43.7	47.6	12.3	15,000
Gates Learjet 35A/36A	D-I	143	39.5	48.7	12.3	18,300
Gates Learjet 54-55-56	C-I	128	43.7	55.1	14.7	21,500

FIGURE B-107 *(continued)* Airplanes arranged by airplane manufacturer, and airport reference code.

Aircraft	Airport Reference Code	Appch Speed Knots	Wingspan Feet	Length Feet	Tail Height Feet	Maximum Takeoff Lbs
General Dynamics 880	D-IV	155	120.0	129.3	36.0	193,500
General Dynamics 990	D-IV	156	120.0	139.2	39.5	255,000
Grumman Gulfstream I	B-II	113	78.3	75.3	23.0	36,000
Grumman Gulfstream II	D-II	141	68.8	79.9	24.5	65,300
Grumman Gulfstream III	C-II	136	77.8	83.1	24.4	68,700
Grumman Gulfstream II-TT	D-II	142	71.7	79.9	24.5	65,300
Grumman Gulfstream IV	D-II	145	77.8	87.8	24.4	71,780
Hamilton Westwind II STD	B-I	96	46.0	45.0	9.2	12,495
HFB-320 Hansa	C-I	125	47.5	54.5	16.2	20,280
Hindustan HS.748-2	B-III	94	98.4	67.0	24.8	44,402
HP Herald	A-III	88	94.8	75.5	24.1	43,000
HS 125 Series 400A	C-I	124	47.0	47.4	16.5	23,300
HS 125 Series 600A	C-I	125	47.0	50.5	17.2	25,000
HS 125 Series 700A	C-I	125	47.0	50.7	17.6	24,200
HS.121 Trident 1E	C-III	137	95.0	114.8	27.0	135,500
HS.121 Trident 2E	C-III	138	98.0	114.8	27.0	144,000
HS.121 Trident 3B	D-III	143	98.0	131.2	28.3	150,000
HS.121 Trident Super 3B	D-III	146	98.0	131.2	28.3	158,000
HS.748 Series 2A	B-III	94	98.5	67.0	24.8	44,490
HS.780 Andover C.Mk.1	B-III	100	98.2	78.0	30.1	50,000
HS.801 Nimrod MR Mk.2	C-III	125 *	114.8	126.8	29.7	177,500
IAI 1121 Jet Comdr.	C-I	130	43.3	50.4	15.8	16,800
IAI Arava-201	A-II	81	68.6	42.7	17.1	15,000
IAI-1124 Westwind	C-I	129	44.8	52.3	15.8	23,500
Ilyushin Il-12	A-III	78	104.0	70.0	30.5	38,000
Ilyushin Il-18	B-IV	103	122.7	117.8	33.3	134,640
Ilyushin Il-62	D-IV	152	141.8	174.3	40.5	363,760
Ilyushin Il-76	B-IV	119	165.7	152.8	48.4	374,785
Ilyushin Il-86	D-IV	141	157.7	195.3	51.8	454,150
Kawasaki C-1	B-III	118 *	100.4	95.1	32.9	85,320
Lapan XT-400	A-I	75	47.9	33.5	14.1	5,555
Learfan 2100	A-I	86	39.3	40.6	12.2	7,400
LET L-410 UVP-E	A-II	81	65.5	47.5	19.1	14,109
Lockheed 100-20 Hercules	C-IV	137	132.6	106.1	39.3	155,000
Lockheed 100-30 Hercules	C-IV	129	132.6	112.7	39.2	155,000
Lockheed 1011-1	C-IV	138	155.3	177.7	55.8	430,000
Lockheed 1011-100	C-IV	140	155.3	177.7	55.8	466,000
Lockheed 1011-200	C-IV	140	155.3	177.7	55.8	466,000
Lockheed 1011-250	D-IV	144	155.3	177.7	55.8	496,000
Lockheed 1011-500	D-IV	144	155.3	164.2	55.8	496,000
Lockheed 1011-500 Ex. Wing	D-IV	148	164.3	164.2	55.8	496,000
Lockheed 1011-600	C-IV	140 *	142.8	141.0	53.0	264,000
Lockheed 1049 Constellat'n	B-IV	113	123.0	113.6	24.8	137,500
Lockheed 1329 JetStar	C-II	132	54.4	60.4	20.4	43,750
Lockheed 1649 Constellat'n	A-IV	89	150.0	116.2	23.4	160,000
Lockheed 188 Electra	C-III	123	99.0	104.6	33.7	116,000
Lockheed 400	C-IV	121 *	119.7	97.8	38.1	84,000
Lockheed 749 Constellat'n	B-IV	93	123.0	95.2	22.4	107,000
Lockheed C-141A Starlifter	C-IV	129	159.9	145.0	39.3	316,600
Lockheed C-141B Starlifter	C-IV	129	159.9	168.3	39.3	343,000
Lockheed C-5B Galaxy	C-VI	135	222.7	247.8	65.1	837,000

FIGURE B-107 (*continued*) Airplanes arranged by airplane manufacturer, and airport reference code.

Aircraft	Airport Reference Code	Appch Speed Knots	Wingspan Feet	Length Feet	Tail Height Feet	Maximum Takeoff Lbs
Lockheed P-3 Orion	C-III	134	99.7	116.8	33.8	135,000
Lockheed SR-71 Blackbird	E-II	180	55.6	107.4	18.5	170,000
MAI-QSTOL	A-III	85	100.3	98.4	32.8	85,300
Marshall (Shorts) Belfast	C-IV	126	158.8	136.4	47.0	230,000
Martin-404	B-III	98	93.3	74.6	28.7	44,900
MDC-C-133	C-V	128	179.7	157.5	48.2	300,000
MDC-DC-10-10	C-IV	136	155.3	182.3	58.4	443,000
MDC-DC-10-30	D-IV	151	165.3	181.6	58.6	590,000
MDC-DC-10-40	D-IV	145	165.4	182.3	58.6	555,000
MDC-DC-3	A-III	72	95.0	64.5	23.5	25,200
MDC-DC-4	B-III	95	117.5	93.9	27.9	73,000
MDC-DC-6A/B	B-III	108	117.5	105.6	29.3	104,000
MDC-DC-7	B-IV	110	127.5	112.3	31.7	143,000
MDC-DC-8-10	C-IV	131	142.4	150.8	43.3	276,000
MDC-DC-8-20/30/40	C-IV	133	142.4	150.8	43.3	315,000
MDC-DC-8-50	C-IV	137	142.4	150.8	43.3	325,000
MDC-DC-8-61	D-IV	142	142.4	187.4	43.0	325,000
MDC-DC-8-62	C-IV	124	148.4	157.5	43.4	350,000
MDC-DC-8-63	D-IV	147	148.4	187.4	43.0	355,000
MDC-DC-9-10/15	C-III	134	89.4	104.4	27.6	90,700
MDC-DC-9-20	C-III	124	93.3	104.4	27.4	98,000
MDC-DC-9-30	C-III	127	93.3	119.3	27.8	110,000
MDC-DC-9-40	C-III	129	93.3	125.6	28.4	114,000
MDC-DC-9-50	C-III	132	93.3	133.6	28.8	121,000
MDC-DC-9-80	C-III	132	107.8	147.8	30.3	140,000
MDC-DC-9-82	C-III	135	107.8	147.8	30.3	149,500
MDC-MD-11	D-IV	155	169.8	201.3	57.8	602,500
Mitsubishi Diamond MU-300	B-I	100	43.5	48.4	13.8	15,730
Mitsubishi Marquise MU-2N	A-I	88	39.2	39.5	13.7	11,575
Mitsubishi MU-2G	B-I	119	39.2	39.5	13.8	10,800
Mitsubishi Solitaire MU-2P	A-I	87	39.2	33.3	12.9	10,470
Nihon YS-11	B-III	98	105.0	86.3	29.5	54,010
Nomad N 22B	A-II	69	54.0	41.2	18.1	8,950
Nomad N 24A	A-II	73	54.2	47.1	18.2	9,400
Partenavia P.68B Victor	A-I	73	39.3	35.6	11.9	6,283
Piaggio PD-808	B-I	117	43.3	42.2	15.8	18,300
Piaggio P-166 Portofino	A-I	82	47.2	39.0	16.4	9,480
Pilatus PC-6 Porter	A-II	57	49.7	37.4	10.5	4,850
Piper 31-310 Navajo	B-I	100	40.7	32.7	13.0	6,200
Piper 400LS Cheyenne	B-I	110	47.7	43.4	17.0	12,050
Piper 60-602P Aerostar	B-I	94	36.7	34.8	12.1	6,000
PZL-AN-2	A-II	54	59.8	41.9	13.1	12,125
PZL-AN-28	A-II	85	72.4	42.9	16.1	14,330
PZL-M-15 Belphegor	A-II	62	73.6	41.9	17.6	12,465
Rockwell 690A Turbo Comdr.	B-I	97	46.5	44.3	14.9	10,300
Rockwell 840	B-II	98	52.1	42.9	14.9	10,325
Rockwell 980	C-II	121	52.1	42.9	14.9	10,325
Rockwell B-1	D-IV	165 *	137.0	147.0	34.0	477,000
Rockwell Sabre 40	B-I	120	44.5	43.8	16.0	18,650
Rockwell Sabre 60	B-I	120	44.5	48.3	16.0	20,000
Rockwell Sabre 65	B-II	105	50.5	46.1	16.0	24,000

FIGURE B-107 *(continued)* Airplanes arranged by airplane manufacturer, and airport reference code.

Aircraft	Airport Reference Code	Appch Speed Knots	Wingspan Feet	Length Feet	Tail Height Feet	Maximum Takeoff Lbs
Rockwell Sabre 75A	C-I	137	44.5	47.2	17.2	23,300
Rockwell Sabre 80	C-II	128	50.4	47.2	17.3	24,500
Shorts 330	B-II	96	74.7	58.0	16.2	22,900
Shorts 360	B-II	104	74.8	70.8	23.7	26,453
Swearingen Merlin 3B	B-I	105	46.2	42.2	16.7	12,500
Swearingen Metro	B-I	112	46.2	59.4	16.7	12,500
Tupolev TU-114	C-IV	132 *	167.6	177.5	50.0	361,620
Tupolev TU-124	C-III	132 *	83.8	100.3	50.0	80,482
Tupolev TU-134	D-III	144	95.2	121.5	30.0	103,600
Tupolev TU-144	E-III	178	94.8	212.6	42.2	396,000
Tupolev TU-154	D-IV	145	123.3	157.2	37.4	216,050
VFW-Fokker 614	B-II	111	70.5	67.5	25.6	44,000
Vickers Vanguard 950	B-IV	119	118.0	122.9	34.9	146,500
Vickers VC-10-1100	C-IV	128	146.2	158.7	39.5	312,000
Vickers VC-10-1150	C-IV	138	146.2	171.7	39.5	335,100
Vickers VC-2-810/840	C-III	122	94.0	85.7	26.8	72,500
Volpar Turbo 18	B-I	100	46.0	37.4	9.6	10,280
Yakovlev YAK-40	C-III	128 *	82.2	65.9	21.3	35,275
Yakovlev YAK-42	C-III	128 *	112.2	119.3	32.2	117,950
Yunshu-11	A-II	80 *	55.7	39.4	15.1	7,150

* Approach speeds estimated.

FIGURE B-107 *(continued)* Airplanes arranged by airplane manufacturer, and airport reference code.

Section 2. Alphabetical Listing (SI units)

Aircraft	Airport Reference Code	Appch Speed Knots	Wingspan Meters	Length Meters	Tail Height Meters	Maximum Takeoff Kg
Aeritalia G-222	B-III	109	28.6	22.7	9.8	27,987
Aerocom Skyliner	A-II	88	16.5	16.6	5.0	5,670
Aerospatiale C 160 Trans.	C-IV	124	40.0	32.4	11.8	49,258
Aerospatiale NORD-262	B-II	96	21.9	19.3	6.2	10,650
Aerospatiale SE 210 Carav.	C-III	127	34.3	32.0	8.7	52,000
Aerospatiale SN 601 Corv.	B-I	118	12.9	13.8	4.2	6,600
Ahrens AR 404	B-II	98	20.1	16.1	5.8	8,391
AIDC/CAF XC-2	A-III	86	24.9	20.1	7.7	12,474
Airbus A-300-600	C-IV	135	44.8	54.1	16.7	165,000
Airbus A-300-B4	C-IV	132	44.8	53.5	16.9	150,003
Airbus A-310-300	C-IV	125	43.9	46.7	15.9	150,000
Airbus A-320-100	C-III	138	33.9	37.6	11.9	66,000
Air-Metal AM-C 111	B-II	96	19.2	16.8	6.4	8,450
AJI Hustler 400	B-I	98	8.5	10.6	3.0	2,722
Antonov AN-10	C-IV	126	38.0	37.0	9.8	55,111
Antonov AN-12	C-IV	127	38.0	33.2	10.5	55,111
Antonov AN-124	C-VI	124	70.7	68.0	20.2	362,874
Antonov AN-14	A-II	52	22.0	11.3	4.6	3,450
Antonov AN-22	C-V	140 *	64.3	50.9	12.6	226,796
Antonov AN-24	B-III	119	29.2	23.5	8.3	21,004
Antonov AN-26	C-III	121	29.2	23.8	8.6	24,004
Antonov AN-28	A-II	88	22.0	13.0	4.9	5,602
Antonov AN-30	B-III	112	29.4	24.4	8.3	23,151
Antonov AN-72	A-III	89 *	25.8	25.8	8.2	29,937
AW.650 Argosy 220	C-III	123	35.1	26.5	8.2	42,184
AW.660 Argosy C.Mk.1	B-III	113	35.1	27.2	8.2	43,998
BAC 111-200	C-III	129	27.0	28.5	7.5	35,834
BAC 111-300	C-III	128	27.0	28.5	7.5	40,143
BAC 111-400	C-III	137	27.0	28.5	7.5	39,463
BAC 111-475	C-III	135	28.5	28.5	7.5	44,679
BAC 111-500	D-III	144	28.5	32.6	7.5	47,400
BAC/Aerospatiale Concord	D-III	162	25.5	62.6	11.4	185,066
BAe 146-100	B-III	113	26.3	26.2	8.6	33,838
BAe 146-200	B-III	117	26.3	28.6	8.6	40,030
BAe 146-300	C-III	121	26.3	31.8	8.6	47,174
BAe Jetstream 31	B-II	99	15.8	14.4	5.3	6,600
Beech Airliner 1900-C	B-II	120 *	16.6	17.6	4.5	7,530
Beech Airliner C99	B-I	107	14.0	13.6	4.4	5,126
Beech Baron 58	B-I	96	11.5	9.1	3.0	2,495
Beech Baron 58P	B-I	101	11.5	9.1	2.8	2,812
Beech Baron 58TC	B-I	101	11.5	9.1	2.8	2,812
Beech Baron B55	A-I	90	11.5	8.5	2.8	2,313
Beech Baron E55	A-I	88	11.5	8.8	2.8	2,404
Beech Bonanza A36	A-I	72	10.2	8.4	2.6	1,656
Beech Bonanza B36TC	A-I	75	11.5	8.4	2.6	1,746
Beech Bonanza F33A	A-I	70	10.2	8.1	2.5	1,542
Beech Bonanza V35B	A-I	70	10.2	8.0	2.0	1,542
Beech Duchess 76	A-I	76	11.6	8.8	2.9	1,769
Beech Duke B60	B-I	98	11.9	10.3	3.7	3,073

FIGURE B-107 *(continued)* Airplanes arranged by airplane manufacturer, and airport reference code.

Aircraft	Airport Reference Code	Appch Speed Knots	Wingspan Meters	Length Meters	Tail Height Meters	Maximum Takeoff Kg
Beech E18S	A-II	87	15.1	10.7	2.9	4,218
Beech King Air B100	B-I	111	14.0	12.2	4.7	5,352
Beech King Air C90-1	B-II	100	15.3	10.8	4.3	4,377
Beech King Air F90	B-I	108	14.0	12.1	4.6	4,967
Beech Sierra 200-B24R	A-I	70	10.0	7.8	2.5	1,247
Beech Skipper 77	A-I	63	9.1	7.3	2.1	760
Beech Sundowner 180-C23	A-I	68	10.0	7.8	2.5	1,111
Beech Super King Air B200	B-II	103	16.6	13.4	4.6	5,670
BN-2A Mk.3 Trislander	A-II	65	16.2	13.9	4.3	4,536
Boeing 707-100	C-IV	139	39.9	44.2	12.7	116,727
Boeing 707-200	D-IV	145	39.9	44.2	12.7	116,727
Boeing 707-320	C-IV	139	43.4	46.6	12.9	141,521
Boeing 707-320B	C-IV	136	44.4	46.6	12.8	152,679
Boeing 707-420	C-IV	132	43.4	46.6	12.9	141,521
Boeing 720	C-IV	133	39.9	41.5	12.6	104,009
Boeing 720B	C-IV	137	39.9	41.7	12.6	106,277
Boeing 727-100	C-III	125	32.9	40.6	10.5	76,657
Boeing 727-200	C-III	138	32.9	46.7	10.6	95,028
Boeing 737-100	C-III	137	28.3	28.7	11.3	49,895
Boeing 737-200	C-III	137	28.3	30.5	11.4	52,390
Boeing 737-300	C-III	137	28.9	33.4	11.2	61,235
Boeing 737-400	C-III	139	28.9	36.5	11.2	68,039
Boeing 737-500	C-III	140 *	28.9	31.0	11.2	60,555
Boeing 747-100	D-V	152	59.6	70.7	19.6	272,155
Boeing 747-200	D-V	152	59.6	70.7	19.7	377,842
Boeing 747-300SR	D-V	141	59.6	70.7	19.6	272,155
Boeing 747-400	D-V	154	64.9	70.7	19.6	394,625
Boeing 747-SP	C-V	140	59.6	56.3	20.1	315,700
Boeing 757	C-IV	135	38.0	47.3	13.7	115,666
Boeing 767-200	C-IV	130	47.6	48.5	16.1	142,882
Boeing 767-300	C-IV	130	47.6	55.0	16.0	158,757
Boeing 777	D-IV	145	47.2	55.3	13.7	172,365
Boeing B-52	D-V	141 *	56.4	48.0	12.4	221,353
Boeing C97 Stratocruiser	B-IV	105	43.1	33.6	11.7	66,134
Boeing E-3	C-IV	137	44.5	46.6	12.8	147,418
Boeing E-4 (747-200)	D-V	152	59.6	70.7	19.7	377,842
Boeing YC-14	A-IV	89	39.3	40.1	14.7	97,976
Bristol Brittania 300/310	B-IV	117	43.4	37.9	11.4	83,915
Canadair CL-44	C-IV	123	43.4	41.7	11.7	95,254
Canadair CL-600	C-II	125	18.8	20.8	6.3	18,711
Casa C-207A Azor	B-III	102	27.8	20.8	7.7	16,511
Casa C-212-200 Aviocar	A-II	81	19.0	15.2	6.3	7,700
Cessna Citation I	B-I	108	14.4	13.3	4.4	5,375
Cessna Citation II	B-II	108	15.8	14.4	4.6	6,033
Cessna Citation III	B-II	114	16.3	16.9	5.1	9,979
Cessna-150	A-I	55	10.0	7.3	2.4	726
Cessna-177 Cardinal	A-I	64	10.8	8.3	2.6	1,134
Cessna-402 Businessliner	B-I	95	12.1	11.0	3.5	2,858
Cessna-404 Titan	B-I	92	14.1	12.0	4.0	3,810
Cessna-414 Chancellor	B-I	94	13.4	11.1	3.5	3,078
Cessna-421 Golden Eagle	B-I	96	12.7	11.0	3.5	3,379

FIGURE B-107 (*continued*) Airplanes arranged by airplane manufacturer, and airport reference code.

AIRPORT ENGINEERING   B.91

Aircraft	Airport Reference Code	Appch Speed Knots	Wingspan Meters	Length Meters	Tail Height Meters	Maximum Takeoff Kg
Cessna-441 Conquest	B-II	100	15.0	11.9	4.0	4,502
Convair 240	B-III	107	28.0	22.8	8.2	18,956
Convair 340	B-III	104	32.1	24.8	8.6	22,271
Convair 440	B-III	106	32.1	24.8	8.6	22,271
Convair 580	B-III	107	32.1	24.8	8.9	24,766
Dassault 1150 Atlantic	C-IV	130 *	37.4	31.8	11.3	45,359
Dassault 941	A-II	59	23.4	23.7	9.4	26,490
Dassault FAL-10	B-I	104	13.1	13.9	4.6	8,500
Dassault FAL-20	B-II	107	16.3	17.2	5.3	13,000
Dassault FAL-200	B-II	114	16.3	17.2	5.3	13,903
Dassault FAL-50	B-II	113	18.9	18.5	7.0	17,001
Dassault FAL-900	B-II	100	19.3	20.2	7.6	20,638
Dassault Mercure	B-III	117	30.5	34.8	11.4	56,472
DHC-2 Beaver	A-I	50	14.6	9.2	2.7	2,313
DHC-4 Caribou	A-III	77	29.1	22.1	9.7	12,927
DHC-5D Buffalo	B-III	91	29.3	24.1	8.7	22,317
DHC-6-300 Twin Otter	A-II	75	19.8	15.8	5.9	5,670
DHC-7 Dash 7-100	A-III	83	28.3	24.6	8.0	19,504
DHC-8 Dash 8-300	A-III	90	27.4	25.7	7.5	18,643
DH.104 Dove 8	A-II	84	17.4	11.9	4.1	4,060
DH.106 Comet 4C	B-III	108	35.1	36.0	9.0	73,482
DH.114 Heron 2	A-II	85	21.8	14.8	4.8	6,123
Dornier DO 28D-2	A-II	74	15.5	11.4	3.9	4,017
Dornier LTA	A-II	74 *	17.8	16.6	5.5	6,849
Embraer-110 Bandeirante	B-II	92	15.3	15.1	5.0	5,900
Embraer-121 Xingu	B-I	92	14.4	12.3	4.8	5,670
Embraer-326 Xavante	B-I	102	10.9	10.6	3.7	5,216
Embraer-820 Navajo Chief	A-I	74	12.4	10.5	4.0	3,175
Fairchild C-119	C-III	122	33.3	26.4	8.4	34,927
Fairchild C-121	A-III	88	33.5	23.1	10.4	27,216
Fairchild FH-227 B,D	B-III	105	29.0	25.3	8.4	20,638
Fairchild F-27 A,J	B-III	109	29.0	23.5	8.4	19,051
FMA IA-50 Guarni II	B-II	101	19.5	14.9	5.8	7,121
Fokker F-27-500	B-III	102	29.0	25.1	8.9	20,412
Fokker F-28-1000	B-II	119	23.6	27.4	8.5	29,484
Fokker F-28-2000	B-II	119	23.6	29.6	8.5	29,484
Fokker F-28-3000	C-III	121	25.1	27.4	8.5	33,112
Fokker F-28-4000	C-III	121	25.1	29.6	8.5	33,112
Fokker F-28-6000	B-III	113	25.1	29.6	8.5	33,112
Foxjet ST-600-8	B-I	97	9.6	9.7	3.1	2,064
GAC-100	A-II	86	21.3	20.5	7.6	13,109
Gates Learjet 24	C-I	128	10.9	13.2	3.8	5,897
Gates Learjet 25	C-I	137	10.9	14.5	3.8	6,804
Gates Learjet 28/29	B-I	120	13.3	14.5	3.7	6,804
Gates Learjet 35A/36A	D-I	143	12.0	14.8	3.7	8,301
Gates Learjet 54-55-56	C-I	128	13.3	16.8	4.5	9,752
General Dynamics 880	D-IV	155	36.6	39.4	11.0	87,770
General Dynamics 990	D-IV	156	36.6	42.4	12.0	115,666
Grumman Gulfstream I	B-II	113	23.9	23.0	7.0	16,329
Grumman Gulfstream II	D-II	141	21.0	24.4	7.5	29,620
Grumman Gulfstream III	C-II	136	23.7	25.3	7.4	31,162

FIGURE B-107 *(continued)* Airplanes arranged by airplane manufacturer, and airport reference code.

Aircraft	Airport Reference Code	Appch Speed Knots	Wingspan Meters	Length Meters	Tail Height Meters	Maximum Takeoff Kg
Grumman Gulfstream II-TT	D-II	142	21.9	24.4	7.5	29,620
Grumman Gulfstream IV	D-II	145	23.7	26.8	7.4	32,559
Hamilton Westwind II STD	B-I	96	14.0	13.7	2.8	5,668
HFB-320 Hansa	C-I	125	14.5	16.6	4.9	9,199
Hindustan HS.748-2	B-III	94	30.0	20.4	7.6	20,140
HP Herald	A-III	88	28.9	23.0	7.3	19,504
HS 125 Series 400A	C-I	124	14.3	14.4	5.0	10,569
HS 125 Series 600A	C-I	125	14.3	15.4	5.2	11,340
HS 125 Series 700A	C-I	125	14.3	15.5	5.4	10,977
HS.121 Trident 1E	C-III	137	29.0	35.0	8.2	61,462
HS.121 Trident 2E	C-III	138	29.9	35.0	8.2	65,317
HS.121 Trident 3B	D-III	143	29.9	40.0	8.6	68,039
HS.121 Trident Super 3B	D-III	146	29.9	40.0	8.6	71,668
HS.748 Series 2A	B-III	94	30.0	20.4	7.6	20,180
HS.780 Andover C.Mk.1	B-III	100	29.9	23.8	9.2	22,680
HS.801 Nimrod MR Mk.2	C-III	125 *	35.0	38.6	9.1	80,513
IAI 1121 Jet Comdr.	C-I	130	13.2	15.4	4.8	7,620
IAI Arava-201	A-II	81	20.9	13.0	5.2	6,804
IAI-1124 Westwind	C-I	129	13.7	15.9	4.8	10,659
Ilyushin Il-12	A-III	78	31.7	21.3	9.3	17,237
Ilyushin Il-18	B-IV	103	37.4	35.9	10.1	61,072
Ilyushin Il-62	D-IV	152	43.2	53.1	12.3	164,999
Ilyushin Il-76	B-IV	119	50.5	46.6	14.8	170,000
Ilyushin Il-86	D-IV	141	48.1	59.5	15.8	205,999
Kawasaki C-1	B-III	118 *	30.6	29.0	10.0	38,701
Lapan XT-400	A-I	75	14.6	10.2	4.3	2,520
Learfan 2100	A-I	86	12.0	12.4	3.7	3,357
LET L-410 UVP-E	A-II	81	20.0	14.5	5.8	6,400
Lockheed 100-20 Hercules	C-IV	137	40.4	32.3	12.0	70,307
Lockheed 100-30 Hercules	C-IV	129	40.4	34.4	11.9	70,307
Lockheed 1011-1	C-IV	138	47.3	54.2	17.0	195,045
Lockheed 1011-100	C-IV	140	47.3	54.2	17.0	211,374
Lockheed 1011-200	C-IV	140	47.3	54.2	17.0	211,374
Lockheed 1011-250	D-IV	144	47.3	54.2	17.0	224,982
Lockheed 1011-500	D-IV	144	47.3	50.0	17.0	224,982
Lockheed 1011-500 Ex. Wing	D-IV	148	50.1	50.0	17.0	224,982
Lockheed 1011-600	C-IV	140 *	43.5	43.0	16.2	119,748
Lockheed 1049 Constellat'n	B-IV	113	37.5	34.6	7.6	62,369
Lockheed 1329 JetStar	C-II	132	16.6	18.4	6.2	19,845
Lockheed 1649 Constellat'n	A-IV	89	45.7	35.4	7.1	72,575
Lockheed 188 Electra	C-III	123	30.2	31.9	10.3	52,617
Lockheed 400	C-IV	121 *	36.5	29.8	11.6	38,102
Lockheed 749 Constellat'n	B-IV	93	37.5	29.0	6.8	48,534
Lockheed C-141A Starlifter	C-IV	129	48.7	44.2	12.0	143,607
Lockheed C-141B Starlifter	C-IV	129	48.7	51.3	12.0	155,582
Lockheed C-5B Galaxy	C-VI	135	67.9	75.5	19.8	379,657
Lockheed P-3 Orion	C-III	134	30.4	35.6	10.3	61,235
Lockheed SR-71 Blackbird	E-II	180	16.9	32.7	5.6	77,111
MAI-QSTOL	A-III	85	30.6	30.0	10.0	38,691
Marshall (Shorts) Belfast	C-IV	126	48.4	41.6	14.3	104,326
Martin-404	B-III	98	28.4	22.7	8.7	20,366

FIGURE B-107 *(continued)* Airplanes arranged by airplane manufacturer, and airport reference code.

Aircraft	Airport Reference Code	Appch Speed Knots	Wingspan Meters	Length Meters	Tail Height Meters	Maximum Takeoff Kg
MDC-C-133	C-V	128	54.8	48.0	14.7	136,078
MDC-DC-10-10	C-IV	136	47.3	55.6	17.8	200,941
MDC-DC-10-30	D-IV	151	50.4	55.4	17.9	267,619
MDC-DC-10-40	D-IV	145	50.4	55.6	17.9	251,744
MDC-DC-3	A-III	72	29.0	19.7	7.2	11,431
MDC-DC-4	B-III	95	35.8	28.6	8.5	33,112
MDC-DC-6A/B	B-III	108	35.8	32.2	8.9	47,174
MDC-DC-7	B-IV	110	38.9	34.2	9.7	64,864
MDC-DC-8-10	C-IV	131	43.4	46.0	13.2	125,191
MDC-DC-8-20/30/40	C-IV	133	43.4	46.0	13.2	142,882
MDC-DC-8-50	C-IV	137	43.4	46.0	13.2	147,418
MDC-DC-8-61	D-IV	142	43.4	57.1	13.1	147,418
MDC-DC-8-62	C-IV	124	45.2	48.0	13.2	158,757
MDC-DC-8-63	D-IV	147	45.2	57.1	13.1	161,025
MDC-DC-9-10/15	C-III	134	27.2	31.8	8.4	41,141
MDC-DC-9-20	C-III	124	28.4	31.8	8.4	44,452
MDC-DC-9-30	C-III	127	28.4	36.4	8.5	49,895
MDC-DC-9-40	C-III	129	28.4	38.3	8.7	51,710
MDC-DC-9-50	C-III	132	28.4	40.7	8.8	54,885
MDC-DC-9-80	C-III	132	32.9	45.0	9.2	63,503
MDC-DC-9-82	C-III	135	32.9	45.0	9.2	67,812
MDC-MD-11	D-IV	155	51.8	61.4	17.6	273,289
Mitsubishi Diamond MU-300	B-I	100	13.3	14.8	4.2	7,135
Mitsubishi Marquise MU-2N	A-I	88	11.9	12.0	4.2	5,250
Mitsubishi MU-2G	B-I	119	11.9	12.0	4.2	4,899
Mitsubishi Solitaire MU-2P	A-I	87	11.9	10.1	3.9	4,749
Nihon YS-11	B-III	98	32.0	26.3	9.0	24,499
Nomad N 22B	A-II	69	16.5	12.6	5.5	4,060
Nomad N 24A	A-II	73	16.5	14.4	5.5	4,264
Partenavia P.68B Victor	A-I	73	12.0	10.9	3.6	2,850
Piaggio PD-808	B-I	117	13.2	12.9	4.8	8,301
Piaggio P-166 Portofino	A-I	82	14.4	11.9	5.0	4,300
Pilatus PC-6 Porter	A-II	57	15.1	11.4	3.2	2,200
Piper 31-310 Navajo	B-I	100	12.4	10.0	4.0	2,812
Piper 400LS Cheyenne	B-I	110	14.5	13.2	5.2	5,466
Piper 60-602P Aerostar	B-I	94	11.2	10.6	3.7	2,722
PZL-AN-2	A-II	54	18.2	12.8	4.0	5,500
PZL-AN-28	A-II	85	22.1	13.1	4.9	6,500
PZL-M-15 Belphegor	A-II	62	22.4	12.8	5.4	5,654
Rockwell 690A Turbo Comdr.	B-I	97	14.2	13.5	4.5	4,672
Rockwell 840	B-II	98	15.9	13.1	4.5	4,683
Rockwell 980	C-II	121	15.9	13.1	4.5	4,683
Rockwell B-1	D-IV	165 *	41.8	44.8	10.4	216,364
Rockwell Sabre 40	B-I	120	13.6	13.4	4.9	8,459
Rockwell Sabre 60	B-I	120	13.6	14.7	4.9	9,072
Rockwell Sabre 65	B-II	105	15.4	14.1	4.9	10,886
Rockwell Sabre 75A	C-I	137	13.6	14.4	5.2	10,569
Rockwell Sabre 80	C-II	128	15.4	14.4	5.3	11,113
Shorts 330	B-II	96	22.8	17.7	4.9	10,387
Shorts 360	B-II	104	22.8	21.6	7.2	11,999
Swearingen Merlin 3B	B-I	105	14.1	12.9	5.1	5,670

FIGURE B-107 *(continued)* Airplanes arranged by airplane manufacturer, and airport reference code.

Aircraft	Airport Reference Code	Appch Speed Knots	Wingspan Meters	Length Meters	Tail Height Meters	Maximum Takeoff Kg
Swearingen Metro	B-I	112	14.1	18.1	5.1	5,670
Tupolev TU-114	C-IV	132 *	51.1	54.1	15.2	164,028
Tupolev TU-124	C-III	132 *	25.5	30.6	15.2	36,506
Tupolev TU-134	D-III	144	29.0	37.0	9.1	46,992
Tupolev TU-144	E-III	178	28.9	64.8	12.9	179,623
Tupolev TU-154	D-IV	145	37.6	47.9	11.4	97,999
VFW-Fokker 614	B-II	111	21.5	20.6	7.8	19,958
Vickers Vanguard 950	B-IV	119	36.0	37.5	10.6	66,451
Vickers VC-10-1100	C-IV	128	44.6	48.4	12.0	141,521
Vickers VC-10-1150	C-IV	138	44.6	52.3	12.0	151,999
Vickers VC-2-810/840	C-III	122	28.7	26.1	8.2	32,885
Volpar Turbo 18	B-I	100	14.0	11.4	2.9	4,663
Yakovlev YAK-40	C-III	128 *	25.1	20.1	6.5	16,000
Yakovlev YAK-42	C-III	128 *	34.2	36.4	9.8	53,501
Yunshu-11	A-II	80 *	17.0	12.0	4.6	3,243

* Approach speeds estimated.

FIGURE B-107 *(continued)* Airplanes arranged by airplane manufacturer, and airport reference code.

Section 3. Listing Small Airplanes by Airport Reference Code (U.S. customary units)

Aircraft	Airport Reference Code	Appch Speed Knots	Wingspan Feet	Length Feet	Tail Height Feet	Maximum Takeoff Lbs
Beech Baron B55	A-I	90	37.8	28.0	9.1	5,100
Beech Baron E55	A-I	88	37.8	29.0	9.1	5,300
Beech Bonanza A36	A-I	72	33.5	27.5	8.6	3,650
Beech Bonanza B36TC	A-I	75	37.8	27.5	8.6	3,850
Beech Bonanza F33A	A-I	70	33.5	26.7	8.2	3,400
Beech Bonanza V35B	A-I	70	33.5	26.4	6.6	3,400
Beech Duchess 76	A-I	76	38.0	29.0	9.5	3,900
Beech Sierra 200-B24R	A-I	70	32.8	25.7	8.2	2,750
Beech Skipper 77	A-I	63	30.0	24.0	6.9	1,675
Beech Sundowner 180-C23	A-I	68	32.8	25.7	8.2	2,450
Cessna-150	A-I	55	32.7	23.8	8.0	1,600
Cessna-177 Cardinal	A-I	64	35.5	27.2	8.5	2,500
DHC-2 Beaver	A-I	50	48.0	30.3	9.0	5,100
Embraer-820 Navajo Chief	A-I	74	40.7	34.6	13.0	7,000
Lapan XT-400	A-I	75	47.9	33.5	14.1	5,555
Learfan 2100	A-I	86	39.3	40.6	12.2	7,400
Mitsubishi Marquise MU-2N	A-I	88	39.2	39.5	13.7	11,575
Mitsubishi Solitaire MU-2P	A-I	87	39.2	33.3	12.9	10,470
Partenavia P.68B Victor	A-I	73	39.3	35.6	11.9	6,283
Piaggio P-166 Portofino	A-I	82	47.2	39.0	16.4	9,480
AJI Hustler 400	B-I	98	28.0	34.8	9.8	6,000
Beech Airliner C99	B-I	107	45.9	44.6	14.4	11,300
Beech Baron 58	B-I	96	37.8	29.8	9.8	5,500
Beech Baron 58P	B-I	101	37.8	29.8	9.1	6,200
Beech Baron 58TC	B-I	101	37.8	29.8	9.1	6,200
Beech Duke B60	B-I	98	39.2	33.8	12.3	6,775
Beech King Air B100	B-I	111	45.8	39.9	15.3	11,800
Beech King Air F90	B-I	108	45.9	39.8	15.1	10,950
Cessna Citation I	B-I	108	47.1	43.5	14.3	11,850
Cessna-402 Businessliner	B-I	95	39.8	36.1	11.6	6,300
Cessna-404 Titan	B-I	92	46.3	39.5	13.2	8,400
Cessna-414 Chancellor	B-I	94	44.1	36.4	11.5	6,785
Cessna-421 Golden Eagle	B-I	96	41.7	36.1	11.6	7,450
Embraer-121 Xingu	B-I	92	47.4	40.2	15.9	12,500
Embraer-326 Xavante	B-I	102	35.6	34.9	12.2	11,500
Foxjet ST-600-8	B-I	97	31.6	31.8	10.2	4,550
Hamilton Westwind II STD	B-I	96	46.0	45.0	9.2	12,495
Mitsubishi MU-2G	B-I	119	39.2	39.5	13.8	10,800
Piper 31-310 Navajo	B-I	100	40.7	32.7	13.0	6,200
Piper 400LS Cheyenne	B-I	110	47.7	43.4	17.0	12,050
Piper 60-602P Aerostar	B-I	94	36.7	34.8	12.1	6,000
Rockwell 690A Turbo Comdr.	B-I	97	46.5	44.3	14.9	10,300
Swearingen Merlin 3B	B-I	105	46.2	42.2	16.7	12,500
Swearingen Metro	B-I	112	46.2	59.4	16.7	12,500
Volpar Turbo 18	B-I	100	46.0	37.4	9.6	10,280
Aerocom Skyliner	A-II	88	54.0	54.3	16.5	12,500
Antonov AN-14	A-II	52	72.1	37.2	15.2	7,607
Antonov AN-28	A-II	88	72.1	42.6	16.1	12,350
Beech E18S	A-II	87	49.7	35.2	9.5	9,300

FIGURE B-107 *(continued)* Airplanes arranged by airplane manufacturer, and airport reference code.

Aircraft	Airport Reference Code	Appch Speed Knots	Wingspan Feet	Length Feet	Tail Height Feet	Maximum Takeoff Lbs
BN-2A Mk.3 Trislander	A-II	65	53.0	45.7	14.2	10,000
DHC-6-300 Twin Otter	A-II	75	65.0	51.7	19.5	12,500
DH.104 Dove 8	A-II	84	57.0	39.2	13.3	8,950
Dornier DO 28D-2	A-II	74	51.0	37.4	12.8	8,855
Nomad N 22B	A-II	69	54.0	41.2	18.1	8,950
Nomad N 24A	A-II	73	54.2	47.1	18.2	9,400
Pilatus PC-6 Porter	A-II	57	49.7	37.4	10.5	4,850
PZL-AN-2	A-II	54	59.8	41.9	13.1	12,125
PZL-M-15 Belphegor	A-II	62	73.6	41.9	17.6	12,465
Yunshu-11	A-II	80 *	55.7	39.4	15.1	7,150
Beech King Air C90-1	B-II	100	50.2	35.5	14.2	9,650
Beech Super King Air B200	B-II	103	54.5	43.8	15.0	12,500
Cessna-441 Conquest	B-II	100	49.3	39.0	13.1	9,925
Rockwell 840	B-II	98	52.1	42.9	14.9	10.325
Rockwell 980	C-II	121	52.1	42.9	14.9	10,325

* Approach speeds estimated.

Section 4. Listing Large Airplanes by Airport Reference Code (U.S. customary units)

Aircraft	Airport Reference Code	Appch Speed Knots	Wingspan Feet	Length Feet	Tail Height Feet	Maximum Takeoff Lbs
Aerospatiale SN 601 Corv.	B-I	118	42.2	45.4	13.9	14,550
Dassault FAL-10	B-I	104	42.9	45.5	15.1	18,740
Gates Learjet 28/29	B-I	120	43.7	47.6	12.3	15,000
Mitsubishi Diamond MU-300	B-I	100	43.5	48.4	13.8	15,730
Piaggio PD-808	B-I	117	43.3	42.2	15.8	18,300
Rockwell Sabre 40	B-I	120	44.5	43.8	16.0	18,650
Rockwell Sabre 60	B-I	120	44.5	48.3	16.0	20,000
Gates Learjet 24	C-I	128	35.6	43.3	12.6	13,000
Gates Learjet 25	C-I	137	35.6	47.6	12.6	15,000
Gates Learjet 54-55-56	C-I	128	43.7	55.1	14.7	21,500
HFB-320 Hansa	C-I	125	47.5	54.5	16.2	20,280
HS 125 Series 400A	C-I	124	47.0	47.4	16.5	23,300
HS 125 Series 600A	C-I	125	47.0	50.5	17.2	25,000
HS 125 Series 700A	C-I	125	47.0	50.7	17.6	24,200
IAI 1121 Jet Comdr.	C-I	130	43.3	50.4	15.8	16,800
IAI-1124 Westwind	C-I	129	44.8	52.3	15.8	23,500
Rockwell Sabre 75A	C-I	137	44.5	47.2	17.2	23.300
Gates Learjet 35A/36A	D-I	143	39.5	48.7	12.3	18.300
Casa C-212-200 Aviocar	A-II	81	62.3	49.8	20.7	16,976
Dassault 941	A-II	59	76.7	77.9	30.7	58,400
DH.114 Heron 2	A-II	85	71.5	48.5	15.6	13,500
Dornier LTA	A-II	74 *	58.4	54.4	18.2	15,100
GAC-100	A-II	86	70.0	67.3	24.9	28,900
IAI Arava-201	A-II	81	68.6	42.7	17.1	15,000
LET L-410 UVP-E	A-II	81	65.5	47.5	19.1	14,109
PZL-AN-28	A-II	85	72.4	42.9	16.1	14.330

FIGURE B-107 *(continued)* Airplanes arranged by airplane manufacturer, and airport reference code.

Aircraft	Airport Reference Code	Appch Speed Knots	Wingspan Feet	Length Feet	Tail Height Feet	Maximum Takeoff Lbs
Aerospatiale NORD-262	B-II	96	71.9	63.3	20.4	23,480
Ahrens AR 404	B-II	98	66.0	52.7	19.0	18,500
Air-Metal AM-C 111	B-II	96	63.0	55.2	21.0	18,629
BAe Jetstream 31	B-II	99	52.0	47.2	17.5	14,550
Beech Airliner 1900-C	B-II	120 *	54.5	57.8	14.9	16,600
Cessna Citation II	B-II	108	51.7	47.2	15.0	13,300
Cessna Citation III	B-II	114	53.5	55.5	16.8	22,000
Dassault FAL-20	B-II	107	53.5	56.3	17.4	28,660
Dassault FAL-200	B-II	114	53.5	56.3	17.4	30,650
Dassault FAL-50	B-II	113	61.9	60.8	22.9	37,480
Dassault FAL-900	B-II	100	63.4	66.3	24.8	45,500
Embraer-110 Bandeirante	B-II	92	50.3	49.5	16.5	13,007
FMA IA-50 Guarni II	B-II	101	64.1	48.8	19.1	15,700
Fokker F-28-1000	B-II	119	77.3	89.9	27.8	65,000
Fokker F-28-2000	B-II	119	77.3	97.2	27.8	65,000
Grumman Gulfstream I	B-II	113	78.3	75.3	23.0	36,000
Rockwell Sabre 65	B-II	105	50.5	46.1	16.0	24,000
Shorts 330	B-II	96	74.7	58.0	16.2	22,900
Shorts 360	B-II	104	74.8	70.8	23.7	26,453
VFW-Fokker 614	B-II	111	70.5	67.5	25.6	44,000
Canadair CL-600	C-II	125	61.8	68.4	20.7	41,250
Grumman Gulfstream III	C-II	136	77.8	83.1	24.4	68,700
Lockheed 1329 JetStar	C-II	132	54.4	60.4	20.4	43,750
Rockwell Sabre 80	C-II	128	50.4	47.2	17.3	24,500
Grumman Gulfstream II	D-II	141	68.8	79.9	24.5	65,300
Grumman Gulfstream II-TT	D-II	142	71.7	79.9	24.5	65,300
Grumman Gulfstream IV	D-II	145	77.8	87.8	24.4	71,780
Lockheed SR-71 Blackbird	E-II	180	55.6	107.4	18.5	170,000
AIDC/CAF XC-2	A-III	86	81.7	65.9	25.3	27,500
Antonov AN-72	A-III	89 *	84.7	84.7	27.0	66,000
DHC-4 Caribou	A-III	77	95.6	72.6	31.8	28,500
DHC-7 Dash 7-100	A-III	83	93.0	80.7	26.2	43,000
DHC-8 Dash 8-300	A-III	90	90.0	84.3	24.6	41,100
Fairchild C-121	A-III	88	110.0	75.8	34.1	60,000
HP Herald	A-III	88	94.8	75.5	24.1	43,000
Ilyushin Il-12	A-III	78	104.0	70.0	30.5	38,000
MAI-QSTOL	A-III	85	100.3	98.4	32.8	85,300
MDC-DC-3	A-III	72	95.0	64.5	23.5	25,200
Aeritalia G-222	B-III	109	93.8	74.4	32.0	61,700
Antonov AN-24	B-III	119	95.8	77.2	27.3	46,305
Antonov AN-30	B-III	112	96.4	80.1	27.3	51,040
AW.660 Argosy C.Mk.1	B-III	113	115.0	89.1	27.0	97,000
BAe 146-100	B-III	113	86.4	85.8	28.3	74,600
BAe 146-200	B-III	117	86.4	93.7	28.3	88,250
Casa C-207A Azor	B-III	102	91.2	68.4	25.4	36,400
Convair 240	B-III	107	91.8	74.7	26.9	41,790
Convair 340	B-III	104	105.3	81.5	28.2	49,100
Convair 440	B-III	106	105.3	81.5	28.2	49,100
Convair 580	B-III	107	105.3	81.5	29.2	54,600
Dassault Mercure	B-III	117	100.2	114.3	37.3	124,500
DHC-5D Buffalo	B-III	91	96.0	79.0	28.7	49,200

FIGURE B-107 *(continued)* Airplanes arranged by airplane manufacturer, and airport reference code.

B.98   APPENDIX B

Aircraft	Airport Reference Code	Appch Speed Knots	Wingspan Feet	Length Feet	Tail Height Feet	Maximum Takeoff Lbs
DH.106 Comet 4C	B-III	108	115.0	118.0	29.5	162,000
Fairchild FH-227 B,D	B-III	105	95.2	83.1	27.5	45,500
Fairchild F-27 A,J	B-III	109	95.2	77.2	27.5	42,000
Fokker F-27-500	B-III	102	95.2	82.3	29.3	45,000
Fokker F-28-6000	B-III	113	82.3	97.2	27.8	73,000
Hindustan HS.748-2	B-III	94	98.4	67.0	24.8	44,402
HS.748 Series 2A	B-III	94	98.5	67.0	24.8	44,490
HS.780 Andover C.Mk.1	B-III	100	98.2	78.0	30.1	50,000
Kawasaki C-1	B-III	118 *	100.4	95.1	32.9	85,320
Martin-404	B-III	98	93.3	74.6	28.7	44,900
MDC-DC-4	B-III	95	117.5	93.9	27.9	73,000
MDC-DC-6A/B	B-III	108	117.5	105.6	29.3	104,000
Nihon YS-11	B-III	98	105.0	86.3	29.5	54,010
Aerospatiale SE 210 Carav.	C-III	127	112.5	105.0	28.6	114,640
Airbus A-320-100	C-III	138	111.3	123.3	39.1	145,505
Antonov AN-26	C-III	121	95.8	78.1	28.1	52,920
AW.650 Argosy 220	C-III	123	115.0	86.8	27.0	93,000
BAC 111-200	C-III	129	88.5	93.5	24.5	79,000
BAC 111-300	C-III	128	88.5	93.5	24.5	88,500
BAC 111-400	C-III	137	88.5	93.5	24.5	87,000
BAC 111-475	C-III	135	93.5	93.5	24.5	98,500
BAe 146-300	C-III	121	86.4	104.2	28.1	104,000
Boeing 727-100	C-III	125	108.0	133.2	34.3	169,000
Boeing 727-200	C-III	138	108.0	153.2	34.9	209,500
Boeing 737-100	C-III	137	93.0	94.0	37.2	110,000
Boeing 737-200	C-III	137	93.0	100.2	37.3	115,500
Boeing 737-300	C-III	137	94.8	109.6	36.6	135,000
Boeing 737-400	C-III	139	94.8	119.6	36.6	150,000
Boeing 737-500	C-III	140 *	94.8	101.8	36.6	133,500
Fairchild C-119	C-III	122	109.3	86.5	27.5	77,000
Fokker F-28-3000	C-III	121	82.3	89.9	27.8	73,000
Fokker F-28-4000	C-III	121	82.3	97.2	27.8	73,000
HS.121 Trident 1E	C-III	137	95.0	114.8	27.0	135,500
HS.121 Trident 2E	C-III	138	98.0	114.8	27.0	144,000
HS.801 Nimrod MR Mk.2	C-III	125 *	114.8	126.8	29.7	177,500
Lockheed 188 Electra	C-III	123	99.0	104.6	33.7	116,000
Lockheed P-3 Orion	C-III	134	99.7	116.8	33.8	135,000
MDC-DC-9-10/15	C-III	134	89.4	104.4	27.6	90,700
MDC-DC-9-20	C-III	124	93.3	104.4	27.4	98,000
MDC-DC-9-30	C-III	127	93.3	119.3	27.8	110,000
MDC-DC-9-40	C-III	129	93.3	125.6	28.4	114,000
MDC-DC-9-50	C-III	132	93.3	133.6	28.8	121,000
MDC-DC-9-80	C-III	132	107.8	147.8	30.3	140,000
MDC-DC-9-82	C-III	135	107.8	147.8	30.3	149,500
Tupolev TU-124	C-III	132 *	83.8	100.3	50.0	80,482
Vickers VC-2-810/840	C-III	122	94.0	85.7	26.8	72,500
Yakovlev YAK-40	C-III	128 *	82.2	65.9	21.3	35,275
Yakovlev YAK-42	C-III	128 *	112.2	119.3	32.2	117,950
BAC 111-500	D-III	144	93.5	107.0	24.5	104,500
BAC/Aerospatiale Concord	D-III	162	83.8	205.4	37.4	408,000
HS.121 Trident 3B	D-III	143	98.0	131.2	28.3	150,000

FIGURE B-107 *(continued)*   Airplanes arranged by airplane manufacturer, and airport reference code.

Aircraft	Airport Reference Code	Appch Speed Knots	Wingspan Feet	Length Feet	Tail Height Feet	Maximum Takeoff Lbs
HS.121 Trident Super 3B	D-III	146	98.0	131.2	28.3	158,000
Tupolev TU-134	D-III	144	95.2	121.5	30.0	103,600
Tupolev TU-144	E-III	178	94.8	212.6	42.2	396,000
Boeing YC-14	A-IV	89	129.0	131.7	48.3	216,000
Lockheed 1649 Constellat'n	A-IV	89	150.0	116.2	23.4	160,000
Boeing C97 Stratocruiser	B-IV	105	141.3	110.3	38.3	145,800
Bristol Brittania 300/310	B-IV	117	142.3	124.2	37.5	185,000
Ilyushin Il-18	B-IV	103	122.7	117.8	33.3	134,640
Ilyushin Il-76	B-IV	119	165.7	152.8	48.4	374,785
Lockheed 1049 Constellat'n	B-IV	113	123.0	113.6	24.8	137,500
Lockheed 749 Constellat'n	B-IV	93	123.0	95.2	22.4	107,000
MDC-DC-7	B-IV	110	127.5	112.3	31.7	143,000
Vickers Vanguard 950	B-IV	119	118.0	122.9	34.9	146,500
Aerospatiale C 160 Trans.	C-IV	124	131.3	106.3	38.7	108,596
Airbus A-300-600	C-IV	135	147.1	177.5	54.7	363,763
Airbus A-300-B4	C-IV	132	147.1	175.5	55.5	330,700
Airbus A-310-300	C-IV	125	144.1	153.2	52.3	330,693
Antonov AN-10	C-IV	126	124.8	121.4	32.2	121,500
Antonov AN-12	C-IV	127	124.8	109.0	34.6	121,500
Boeing 707-100	C-IV	139	130.8	145.1	41.7	257,340
Boeing 707-320	C-IV	139	142.4	152.9	42.2	312,000
Boeing 707-320B	C-IV	136	145.8	152.9	42.1	336,600
Boeing 707-420	C-IV	132	142.4	152.9	42.2	312,000
Boeing 720	C-IV	133	130.8	136.2	41.4	229,300
Boeing 720B	C-IV	137	130.8	136.8	41.2	234,300
Boeing 757	C-IV	135	124.8	155.3	45.1	255,000
Boeing 767-200	C-IV	130	156.1	159.2	52.9	315,000
Boeing 767-300	C-IV	130	156.1	180.3	52.6	350,000
Boeing E-3	C-IV	137	145.9	153.0	42.0	325,000
Canadair CL-44	C-IV	123	142.3	136.8	38.4	210,000
Dassault 1150 Atlantic	C-IV	130 *	122.7	104.2	37.2	100,000
Lockheed 100-20 Hercules	C-IV	137	132.6	106.1	39.3	155,000
Lockheed 100-30 Hercules	C-IV	129	132.6	112.7	39.2	155,000
Lockheed 1011-1	C-IV	138	155.3	177.7	55.8	430,000
Lockheed 1011-100	C-IV	140	155.3	177.7	55.8	466,000
Lockheed 1011-200	C-IV	140	155.3	177.7	55.8	466,000
Lockheed 1011-600	C-IV	140 *	142.8	141.0	53.0	264,000
Lockheed 400	C-IV	121 *	119.7	97.8	38.1	84,000
Lockheed C-141A Starlifter	C-IV	129	159.9	145.0	39.3	316,600
Lockheed C-141B Starlifter	C-IV	129	159.9	168.3	39.3	343,000
Marshall (Shorts) Belfast	C-IV	126	158.8	136.4	47.0	230,000
MDC-DC-10-10	C-IV	136	155.3	182.3	58.4	443,000
MDC-DC-8-10	C-IV	131	142.4	150.8	43.3	276,000
MDC-DC-8-20/30/40	C-IV	133	142.4	150.8	43.3	315,000
MDC-DC-8-50	C-IV	137	142.4	150.8	43.3	325,000
MDC-DC-8-62	C-IV	124	148.4	157.5	43.4	350,000
Tupolev TU-114	C-IV	132 *	167.6	177.5	50.0	361,620
Vickers VC-10-1100	C-IV	128	146.2	158.7	39.5	312,000
Vickers VC-10-1150	C-IV	138	146.2	171.7	39.5	335,100
Boeing 707-200	D-IV	145	130.8	145.1	41.7	257,340
Boeing 777	D-IV	145	155.0	181.5	44.8	380,000

FIGURE B-107 *(continued)* Airplanes arranged by airplane manufacturer, and airport reference code.

B.100 APPENDIX B

Aircraft	Airport Reference Code	Appch Speed Knots	Wingspan Feet	Length Feet	Tail Height Feet	Maximum Takeoff Lbs
General Dynamics 880	D-IV	155	120.0	129.3	36.0	193,500
General Dynamics 990	D-IV	156	120.0	139.2	39.5	255,000
Ilyushin Il-62	D-IV	152	141.8	174.3	40.5	363,760
Ilyushin Il-86	D-IV	141	157.7	195.3	51.8	454,150
Lockheed 1011-250	D-IV	144	155.3	177.7	55.8	496,000
Lockheed 1011-500	D-IV	144	155.3	164.2	55.8	496,000
Lockheed 1011-500 Ex. Wing	D-IV	148	164.3	164.2	55.8	496,000
MDC-DC-10-30	D-IV	151	165.3	181.6	58.6	590,000
MDC-DC-10-40	D-IV	145	165.4	182.3	58.6	555,000
MDC-DC-8-61	D-IV	142	142.4	187.4	43.0	325,000
MDC-DC-8-63	D-IV	147	148.4	187.4	43.0	355,000
MDC-MD-11	D-IV	155	169.8	201.3	57.8	602,500
Rockwell B-1	D-IV	165 *	137.0	147.0	34.0	477,000
Tupolev TU-154	D-IV	145	123.3	157.2	37.4	216,050
Antonov AN-22	C-V	140 *	211.0	167.0	41.2	500,000
Boeing 747-SP	C-V	140	195.7	184.8	65.8	696,000
MDC-C-133	C-V	128	179.7	157.5	48.2	300,000
Boeing 747-100	D-V	152	195.7	231.8	64.3	600,000
Boeing 747-200	D-V	152	195.7	231.8	64.7	833,000
Boeing 747-300SR	D-V	141	195.7	231.8	64.3	600,000
Boeing 747-400	D-V	154	213.0	231.8	64.3	870,000
Boeing B-52	D-V	141 *	185.0	157.6	40.8	488,000
Boeing E-4 (747-200)	D-V	152	195.7	231.8	64.7	833,000
Antonov AN-124	C-VI	124	232.0	223.0	66.2	800,000
Lockheed C-5B Galaxy	C-VI	135	222.7	247.8	65.1	837,000

* Approach speeds estimated.

FIGURE B-107 *(continued)* Airplanes arranged by airplane manufacturer, and airport reference code.

Section 5. Listing Small Airplanes by Airport Reference Code (SI units)

Aircraft	Airport Reference Code	Appch Speed Knots	Wingspan Meters	Length Meters	Tail Height Meters	Maximum Takeoff Kg
Beech Baron B55	A-I	90	11.5	8.5	2.8	2,313
Beech Baron E55	A-I	88	11.5	8.8	2.8	2,404
Beech Bonanza A36	A-I	72	10.2	8.4	2.6	1,656
Beech Bonanza B36TC	A-I	75	11.5	8.4	2.6	1,746
Beech Bonanza F33A	A-I	70	10.2	8.1	2.5	1,542
Beech Bonanza V35B	A-I	70	10.2	8.0	2.0	1,542
Beech Duchess 76	A-I	76	11.6	8.8	2.9	1,769
Beech Sierra 200-B24R	A-I	70	10.0	7.8	2.5	1,247
Beech Skipper 77	A-I	63	9.1	7.3	2.1	760
Beech Sundowner 180-C23	A-I	68	10.0	7.8	2.5	1,111
Cessna-150	A-I	55	10.0	7.3	2.4	726
Cessna-177 Cardinal	A-I	64	10.8	8.3	2.6	1,134
DHC-2 Beaver	A-I	50	14.6	9.2	2.7	2,313
Embraer-820 Navajo Chief	A-I	74	12.4	10.5	4.0	3,175
Lapan XT-400	A-I	75	14.6	10.2	4.3	2,520
Learfan 2100	A-I	86	12.0	12.4	3.7	3,357
Mitsubishi Marquise MU-2N	A-I	88	11.9	12.0	4.2	5,250
Mitsubishi Solitaire MU-2P	A-I	87	11.9	10.1	3.9	4,749
Partenavia P.68B Victor	A-I	73	12.0	10.9	3.6	2,850
Piaggio P-166 Portofino	A-I	82	14.4	11.9	5.0	4,300
AJI Hustler 400	B-I	98	8.5	10.6	3.0	2,722
Beech Airliner C99	B-I	107	14.0	13.6	4.4	5,126
Beech Baron 58	B-I	96	11.5	9.1	3.0	2,495
Beech Baron 58P	B-I	101	11.5	9.1	2.8	2,812
Beech Baron 58TC	B-I	101	11.5	9.1	2.8	2,812
Beech Duke B60	B-I	98	11.9	10.3	3.7	3,073
Beech King Air B100	B-I	111	14.0	12.2	4.7	5,352
Beech King Air F90	B-I	108	14.0	12.1	4.6	4,967
Cessna Citation I	B-I	108	14.4	13.3	4.4	5,375
Cessna-402 Businessliner	B-I	95	12.1	11.0	3.5	2,858
Cessna-404 Titan	B-I	92	14.1	12.0	4.0	3,810
Cessna-414 Chancellor	B-I	94	13.4	11.1	3.5	3,078
Cessna-421 Golden Eagle	B-I	96	12.7	11.0	3.5	3,379
Embraer-121 Xingu	B-I	92	14.4	12.3	4.8	5,670
Embraer-326 Xavante	B-I	102	10.9	10.6	3.7	5,216
Foxjet ST-600-8	B-I	97	9.6	9.7	3.1	2,064
Hamilton Westwind II STD	B-I	96	14.0	13.7	2.8	5,668
Mitsubishi MU-2G	B-I	119	11.9	12.0	4.2	4,899
Piper 31-310 Navajo	B-I	100	12.4	10.0	4.0	2,812
Piper 400LS Cheyenne	B-I	110	14.5	13.2	5.2	5,466
Piper 60-602P Aerostar	B-I	94	11.2	10.6	3.7	2,722
Rockwell 690A Turbo Comdr.	B-I	97	14.2	13.5	4.5	4,672
Swearingen Merlin 3B	B-I	105	14.1	12.9	5.1	5,670
Swearingen Metro	B-I	112	14.1	18.1	5.1	5,670
Volpar Turbo 18	B-I	100	14.0	11.4	2.9	4,663
Aerocom Skyliner	A-II	88	16.5	16.6	5.0	5,670
Antonov AN-14	A-II	52	22.0	11.3	4.6	3,450
Antonov AN-28	A-II	88	22.0	13.0	4.9	5,602
Beech E18S	A-II	87	15.1	10.7	2.9	4,218

FIGURE B-107 *(continued)* Airplanes arranged by airplane manufacturer, and airport reference code.

Aircraft	Airport Reference Code	Appch Speed Knots	Wingspan Meters	Length Meters	Tail Height Meters	Maximum Takeoff Kg
BN-2A Mk.3 Trislander	A-II	65	16.2	13.9	4.3	4,536
DHC-6-300 Twin Otter	A-II	75	19.8	15.8	5.9	5,670
DH.104 Dove 8	A-II	84	17.4	11.9	4.1	4,060
Dornier DO 28D-2	A-II	74	15.5	11.4	3.9	4,017
Nomad N 22B	A-II	69	16.5	12.6	5.5	4,060
Nomad N 24A	A-II	73	16.5	14.4	5.5	4,264
Pilatus PC-6 Porter	A-II	57	15.1	11.4	3.2	2,200
PZL-AN-2	A-II	54	18.2	12.8	4.0	5,500
PZL-M-15 Belphegor	A-II	62	22.4	12.8	5.4	5,654
Yunshu-11	A-II	80 *	17.0	12.0	4.6	3,243
Beech King Air C90-1	B-II	100	15.3	10.8	4.3	4,377
Beech Super King Air B200	B-II	103	16.6	13.4	4.6	5,670
Cessna-441 Conquest	B-II	100	15.0	11.9	4.0	4,502
Rockwell 840	B-II	98	15.9	13.1	4.5	4,683
Rockwell 980	C-II	121	15.9	13.1	4.5	4,683

* Approach speeds estimated.

Section 6. Listing Large Airplanes by Airport Reference Code (SI units)

Aircraft	Airport Reference Code	Appch Speed Knots	Wingspan Meters	Length Meters	Tail Height Meters	Maximum Takeoff Kg
Aerospatiale SN 601 Corv.	B-I	118	12.9	13.8	4.2	6,600
Dassault FAL-10	B-I	104	13.1	13.9	4.6	8,500
Gates Learjet 28/29	B-I	120	13.3	14.5	3.7	6,804
Mitsubishi Diamond MU-300	B-I	100	13.3	14.8	4.2	7,135
Piaggio PD-808	B-I	117	13.2	12.9	4.8	8,301
Rockwell Sabre 40	B-I	120	13.6	13.4	4.9	8,459
Rockwell Sabre 60	B-I	120	13.6	14.7	4.9	9,072
Gates Learjet 24	C-I	128	10.9	13.2	3.8	5,897
Gates Learjet 25	C-I	137	10.9	14.5	3.8	6,804
Gates Learjet 54-55-56	C-I	128	13.3	16.8	4.5	9,752
HFB-320 Hansa	C-I	125	14.5	16.6	4.9	9,199
HS 125 Series 400A	C-I	124	14.3	14.4	5.0	10,569
HS 125 Series 600A	C-I	125	14.3	15.4	5.2	11,340
HS 125 Series 700A	C-I	125	14.3	15.5	5.4	10,977
IAI-1121 Jet Comdr.	C-I	130	13.2	15.4	4.8	7,620
IAI-1124 Westwind	C-I	129	13.7	15.9	4.8	10,659
Rockwell Sabre 75A	C-I	137	13.6	14.4	5.2	10,569
Gates Learjet 35A/36A	D-I	143	12.0	14.8	3.7	8,301
Casa C-212-200 Aviocar	A-II	81	19.0	15.2	6.3	7,700
Dassault 941	A-II	59	23.4	23.7	9.4	26,490
DH.114 Heron 2	A-II	85	21.8	14.8	4.8	6,123
Dornier LTA	A-II	74 *	17.8	16.6	5.5	6,849
GAC-100	A-II	86	21.3	20.5	7.6	13,109
IAI Arava-201	A-II	81	20.9	13.0	5.2	6,804
LET L-410 UVP-E	A-II	81	20.0	14.5	5.8	6,400
PZL-AN-28	A-II	85	22.1	13.1	4.9	6,500

FIGURE B-107 *(continued)* Airplanes arranged by airplane manufacturer, and airport reference code.

Aircraft	Airport Reference Code	Appch Speed Knots	Wingspan Meters	Length Meters	Tail Height Meters	Maximum Takeoff Kg
Aerospatiale NORD-262	B-II	96	21.9	19.3	6.2	10,650
Ahrens AR 404	B-II	98	20.1	16.1	5.8	8,391
Air-Metal AM-C 111	B-II	96	19.2	16.8	6.4	8,450
BAe Jetstream 31	B-II	99	15.8	14.4	5.3	6,600
Beech Airliner 1900-C	B-II	120 *	16.6	17.6	4.5	7,530
Cessna Citation II	B-II	108	15.8	14.4	4.6	6,033
Cessna Citation III	B-II	114	16.3	16.9	5.1	9,979
Dassault FAL-20	B-II	107	16.3	17.2	5.3	13,000
Dassault FAL-200	B-II	114	16.3	17.2	5.3	13,903
Dassault FAL-50	B-II	113	18.9	18.5	7.0	17,001
Dassault FAL-900	B-II	100	19.3	20.2	7.6	20,638
Embraer-110 Bandeirante	B-II	92	15.3	15.1	5.0	5,900
FMA IA-50 Guarni II	B-II	101	19.5	14.9	5.8	7,121
Fokker F-28-1000	B-II	119	23.6	27.4	8.5	29,484
Fokker F-28-2000	B-II	119	23.6	29.6	8.5	29,484
Grumman Gulfstream I	B-II	113	23.9	23.0	7.0	16,329
Rockwell Sabre 65	B-II	105	15.4	14.1	4.9	10,886
Shorts 330	B-II	96	22.8	17.7	4.9	10,387
Shorts 360	B-II	104	22.8	21.6	7.2	11,999
VFW-Fokker 614	B-II	111	21.5	20.6	7.8	19,958
Canadair CL-600	C-II	125	18.8	20.8	6.3	18,711
Grumman Gulfstream III	C-II	136	23.7	25.3	7.4	31,162
Lockheed 1329 JetStar	C-II	132	16.6	18.4	6.2	19,845
Rockwell Sabre 80	C-II	128	15.4	14.4	5.3	11,113
Grumman Gulfstream II	D-II	141	21.0	24.4	7.5	29,620
Grumman Gulfstream II-TT	D-II	142	21.9	24.4	7.5	29,620
Grumman Gulfstream IV	D-II	145	23.7	26.8	7.4	32,559
Lockheed SR-71 Blackbird	E-II	180	16.9	32.7	5.6	77,111
AIDC/CAF XC-2	A-III	86	24.9	20.1	7.7	12,474
Antonov AN-72	A-III	89 *	25.8	25.8	8.2	29,937
DHC-4 Caribou	A-III	77	29.1	22.1	9.7	12,927
DHC-7 Dash 7-100	A-III	83	28.3	24.6	8.0	19,504
DHC-8 Dash 8-300	A-III	90	27.4	25.7	7.5	18,643
Fairchild C-121	A-III	88	33.5	23.1	10.4	27,216
HP Herald	A-III	88	28.9	23.0	7.3	19,504
Ilyushin Il-12	A-III	78	31.7	21.3	9.3	17,237
MAI-QSTOL	A-III	85	30.6	30.0	10.0	38,691
MDC-DC-3	A-III	72	29.0	19.7	7.2	11,431
Aeritalia G-222	B-III	109	28.6	22.7	9.8	27,987
Antonov AN-24	B-III	119	29.2	23.5	8.3	21,004
Antonov AN-30	B-III	112	29.4	24.4	8.3	23,151
AW.660 Argosy C.Mk.1	B-III	113	35.1	27.2	8.2	43,998
BAe 146-100	B-III	113	26.3	26.2	8.6	33,838
BAe 146-200	B-III	117	26.3	28.6	8.6	40,030
Casa C-207A Azor	B-III	102	27.8	20.8	7.7	16,511
Convair 240	B-III	107	28.0	22.8	8.2	18,956
Convair 340	B-III	104	32.1	24.8	8.6	22,271
Convair 440	B-III	106	32.1	24.8	8.6	22,271
Convair 580	B-III	107	32.1	24.8	8.9	24,766
Dassault Mercure	B-III	117	30.5	34.8	11.4	56,472
DHC-5D Buffalo	B-III	91	29.3	24.1	8.7	22,317

FIGURE B-107 *(continued)* Airplanes arranged by airplane manufacturer, and airport reference code.

Aircraft	Airport Reference Code	Appch Speed Knots	Wingspan Meters	Length Meters	Tail Height Meters	Maximum Takeoff Kg
DH.106 Comet 4C	B-III	108	35.1	36.0	9.0	73,482
Fairchild FH-227 B,D	B-III	105	29.0	25.3	8.4	20,638
Fairchild F-27 A,J	B-III	109	29.0	23.5	8.4	19,051
Fokker F-27-500	B-III	102	29.0	25.1	8.9	20,412
Fokker F-28-6000	B-III	113	25.1	29.6	8.5	33,112
Hindustan HS.748-2	B-III	94	30.0	20.4	7.6	20,140
HS:748 Series 2A	B-III	94	30.0	20.4	7.6	20,180
HS.780 Andover C.Mk.1	B-III	100	29.9	23.8	9.2	22,680
Kawasaki C-1	B-III	118 *	30.6	29.0	10.0	38,701
Martin-404	B-III	98	28.4	22.7	8.7	20,366
MDC-DC-4	B-III	95	35.8	28.6	8.5	33,112
MDC-DC-6A/B	B-III	108	35.8	32.2	8.9	47,174
Nihon YS-11	B-III	98	32.0	26.3	9.0	24,499
Aerospatiale SE 210 Carav.	C-III	127	34.3	32.0	8.7	52,000
Airbus A-320-100	C-III	138	33.9	37.6	11.9	66,000
Antonov AN-26	C-III	121	29.2	23.8	8.6	24,004
AW.650 Argosy 220	C-III	123	35.1	26.5	8.2	42,184
BAC 111-200	C-III	129	27.0	28.5	7.5	35,834
BAC 111-300	C-III	128	27.0	28.5	7.5	40,143
BAC 111-400	C-III	137	27.0	28.5	7.5	39,463
BAC 111-475	C-III	135	28.5	28.5	7.5	44,679
BAe 146-300	C-III	121	26.3	31.8	8.6	47,174
Boeing 727-100	C-III	125	32.9	40.6	10.5	76,657
Boeing 727-200	C-III	138	32.9	46.7	10.6	95,028
Boeing 737-100	C-III	137	28.3	28.7	11.3	49,895
Boeing 737-200	C-III	137	28.3	30.5	11.4	52,390
Boeing 737-300	C-III	137	28.9	33.4	11.2	61,235
Boeing 737-400	C-III	139	28.9	36.5	11.2	68,039
Boeing 737-500	C-III	140 *	28.9	31.0	11.2	60,555
Fairchild C-119	C-III	122	33.3	26.4	8.4	34,927
Fokker F-28-3000	C-III	121	25.1	27.4	8.5	33,112
Fokker F-28-4000	C-III	121	25.1	29.6	8.5	33,112
HS.121 Trident 1E	C-III	137	29.0	35.0	8.2	61,462
HS.121 Trident 2E	C-III	138	29.9	35.0	8.2	65,317
HS.801 Nimrod MR Mk.2	C-III	125 *	35.0	38.6	9.1	80,513
Lockheed 188 Electra	C-III	123	30.2	31.9	10.3	52,617
Lockheed P-3 Orion	C-III	134	30.4	35.6	10.3	61,235
MDC-DC-9-10/15	C-III	134	27.2	31.8	8.4	41,141
MDC-DC-9-20	C-III	124	28.4	31.8	8.4	44,452
MDC-DC-9-30	C-III	127	28.4	36.4	8.5	49,895
MDC-DC-9-40	C-III	129	28.4	38.3	8.7	51,710
MDC-DC-9-50	C-III	132	28.4	40.7	8.8	54,885
MDC-DC-9-80	C-III	132	32.9	45.0	9.2	63,503
MDC-DC-9-82	C-III	135	32.9	45.0	9.2	67,812
Tupolev TU-124	C-III	132 *	25.5	30.6	15.2	36,506
Vickers VC-2-810/840	C-III	122	28.7	26.1	8.2	32,885
Yakovlev YAK-40	C-III	128 *	25.1	20.1	6.5	16,000
Yakovlev YAK-42	C-III	128 *	34.2	36.4	9.8	53,501
BAC 111-500	D-III	144	28.5	32.6	7.5	47,400
BAC/Aerospatiale Concord	D-III	162	25.5	62.6	11.4	185,066
HS.121 Trident 3B	D-III	143	29.9	40.0	8.6	68,039

FIGURE B-107 *(continued)* Airplanes arranged by airplane manufacturer, and airport reference code.

Aircraft	Airport Reference Code	Appch Speed Knots	Wingspan Meters	Length Meters	Tail Height Meters	Maximum Takeoff Kg
HS.121 Trident Super 3B	D-III	146	29.9	40.0	8.6	71,668
Tupolev TU-134	D-III	144	29.0	37.0	9.1	46,992
Tupolev TU-144	E-III	178	28.9	64.8	12.9	179,623
Boeing YC-14	A-IV	89	39.3	40.1	14.7	97,976
Lockheed 1649 Constellat'n	A-IV	89	45.7	35.4	7.1	72,575
Boeing C97 Stratocruiser	B-IV	105	43.1	33.6	11.7	66,134
Bristol Brittania 300/310	B-IV	117	43.4	37.9	11.4	83,915
Ilyushin Il-18	B-IV	103	37.4	35.9	10.1	61,072
Ilyushin Il-76	B-IV	119	50.5	46.6	14.8	170,000
Lockheed 1049 Constellat'n	B-IV	113	37.5	34.6	7.6	62,369
Lockheed 749 Constellat'n	B-IV	93	37.5	29.0	6.8	48,534
MDC-DC-7	B-IV	110	38.9	34.2	9.7	64,864
Vickers Vanguard 950	B-IV	119	36.0	37.5	10.6	66,451
Aerospatiale C 160 Trans.	C-IV	124	40.0	32.4	11.8	49,258
Airbus A-300-600	C-IV	135	44.8	54.1	16.7	165,000
Airbus A-300-B4	C-IV	132	44.8	53.5	16.9	150,003
Airbus A-310-300	C-IV	125	43.9	46.7	15.9	150,000
Antonov AN-10	C-IV	126	38.0	37.0	9.8	55,111
Antonov AN-12	C-IV	127	38.0	33.2	10.5	55,111
Boeing 707-100	C-IV	139	39.9	44.2	12.7	116,727
Boeing 707-320	C-IV	139	43.4	46.6	12.9	141,521
Boeing 707-320B	C-IV	136	44.4	46.6	12.8	152,679
Boeing 707-420	C-IV	132	43.4	46.6	12.9	141,521
Boeing 720	C-IV	133	39.9	41.5	12.6	104,009
Boeing 720B	C-IV	137	39.9	41.7	12.6	106,277
Boeing 757	C-IV	135	38.0	47.3	13.7	115,666
Boeing 767-200	C-IV	130	47.6	48.5	16.1	142,882
Boeing 767-300	C-IV	130	47.6	55.0	16.0	158,757
Boeing E-3	C-IV	137	44.5	46.6	12.8	147,418
Canadair CL-44	C-IV	123	43.4	41.7	11.7	95,254
Dassault 1150 Atlantic	C-IV	130 *	37.4	31.8	11.3	45,359
Lockheed 100-20 Hercules	C-IV	137	40.4	32.3	12.0	70,307
Lockheed 100-30 Hercules	C-IV	129	40.4	34.4	11.9	70,307
Lockheed 1011-1	C-IV	138	47.3	54.2	17.0	195,045
Lockheed 1011-100	C-IV	140	47.3	54.2	17.0	211,374
Lockheed 1011-200	C-IV	140	47.3	54.2	17.0	211,374
Lockheed 1011-600	C-IV	140 *	43.5	43.0	16.2	119,748
Lockheed 400	C-IV	121 *	36.5	29.8	11.6	38,102
Lockheed C-141A Starlifter	C-IV	129	48.7	44.2	12.0	143,607
Lockheed C-141B Starlifter	C-IV	129	48.7	51.3	12.0	155,582
Marshall (Shorts) Belfast	C-IV	126	48.4	41.6	14.3	104,326
MDC-DC-10-10	C-IV	136	47.3	55.6	17.8	200,941
MDC-DC-8-10	C-IV	131	43.4	46.0	13.2	125,191
MDC-DC-8-20/30/40	C-IV	133	43.4	46.0	13.2	142,882
MDC-DC-8-50	C-IV	137	43.4	46.0	13.2	147,418
MDC-DC-8-62	C-IV	124	45.2	48.0	13.2	158,757
Tupolev TU-114	C-IV	132 *	51.1	54.1	15.2	164,028
Vickers VC-10-1100	C-IV	128	44.6	48.4	12.0	141,521
Vickers VC-10-1150	C-IV	138	44.6	52.3	12.0	151,999
Boeing 707-200	D-IV	145	39.9	44.2	12.7	116,727
Boeing 777	D-IV	145	47.2	55.3	13.7	172,365

FIGURE B-107 *(continued)* Airplanes arranged by airplane manufacturer, and airport reference code.

Aircraft	Airport Reference Code	Appch Speed Knots	Wingspan Meters	Length Meters	Tail Height Meters	Maximum Takeoff Kg
General Dynamics 880	D-IV	155	36.6	39.4	11.0	87,770
General Dynamics 990	D-IV	156	36.6	42.4	12.0	115,666
Ilyushin Il-62	D-IV	152	43.2	53.1	12.3	164,999
Ilyushin Il-86	D-IV	141	48.1	59.5	15.8	205,999
Lockheed 1011-250	D-IV	144	47.3	54.2	17.0	224,982
Lockheed 1011-500	D-IV	144	47.3	50.0	17.0	224,982
Lockheed 1011-500 Ex. Wing	D-IV	148	50.1	50.0	17.0	224,982
MDC-DC-10-30	D-IV	151	50.4	55.4	17.9	267,619
MDC-DC-10-40	D-IV	145	50.4	55.6	17.9	251,744
MDC-DC-8-61	D-IV	142	43.4	57.1	13.1	147,418
MDC-DC-8-63	D-IV	147	45.2	57.1	13.1	161,025
MDC-MD-11	D-IV	155	51.8	61.4	17.6	273,289
Rockwell B-1	D-IV	165 *	41.8	44.8	10.4	216,364
Tupolev TU-154	D-IV	145	37.6	47.9	11.4	97,999
Antonov AN-22	C-V	140 *	64.3	50.9	12.6	226,796
Boeing 747-SP	C-V	140	59.6	56.3	20.1	315,700
MDC-C-133	C-V	128	54.8	48.0	14.7	136,078
Boeing 747-100	D-V	152	59.6	70.7	19.6	272,155
Boeing 747-200	D-V	152	59.6	70.7	19.7	377,842
Boeing 747-300SR	D-V	141	59.6	70.7	19.6	272,155
Boeing 747-400	D-V	154	64.9	70.7	19.6	394,625
Boeing B-52	D-V	141 *	56.4	48.0	12.4	221,353
Boeing E-4 (747-200)	D-V	152	59.6	70.7	19.7	377,842
Antonov AN-124	C-VI	124	70.7	68.0	20.2	362,874
Lockheed C-5B Galaxy	C-VI	135	67.9	75.5	19.8	379,657

* Approach speeds estimated.

FIGURE B-107 *(continued)* Airplanes arranged by airplane manufacturer, and airport reference code.

Aircraft	Airport Reference Code	Appch Speed Knots	Wingspan Meters	Length Meters	Tail Height Meters	Maximum Takeoff Kg
General Dynamics 880	D-IV	155	36.6	39.4	11.0	87,770
General Dynamics 990	D-IV	156	36.6	42.4	12.0	115,666
Ilyushin Il-62	D-IV	152	43.2	53.1	12.3	164,999
Ilyushin Il-86	D-IV	141	48.1	59.5	15.8	205,999
Lockheed 1011-250	D-IV	144	47.3	54.2	17.0	224,982
Lockheed 1011-500	D-IV	144	47.3	50.0	17.0	224,982
Lockheed 1011-500 Ex. Wing	D-IV	148	50.1	50.0	17.0	224,982
MDC-DC-10-30	D-IV	151	50.4	55.4	17.9	267,619
MDC-DC-10-40	D-IV	145	50.4	55.6	17.9	251,744
MDC-DC-8-61	D-IV	142	43.4	57.1	13.1	147,418
MDC-DC-8-63	D-IV	147	45.2	57.1	13.1	161,025
MDC-MD-11	D-IV	155	51.8	61.4	17.6	273,289
Rockwell B-1	D-IV	165 *	41.8	44.8	10.4	216,364
Tupolev TU-154	D-IV	145	37.6	47.9	11.4	97,999
Antonov AN-22	C-V	140 *	64.3	50.9	12.6	226,796
Boeing 747-SP	C-V	140	59.6	56.3	20.1	315,700
MDC-C-133	C-V	128	54.8	48.0	14.7	136,078
Boeing 747-100	D-V	152	59.6	70.7	19.6	272,155
Boeing 747-200	D-V	152	59.6	70.7	19.7	377,842
Boeing 747-300SR	D-V	141	59.6	70.7	19.6	272,155
Boeing 747-400	D-V	154	64.9	70.7	19.6	394,625
Boeing B-52	D-V	141 *	56.4	48.0	12.4	221,353
Boeing E-4 (747-200)	D-V	152	59.6	70.7	19.7	377,842
Antonov AN-124	C-VI	124	70.7	68.0	20.2	362,874
Lockheed C-5B Galaxy	C-VI	135	67.9	75.5	19.8	379,657

* Approach speeds estimated.

FIGURE B-107 *(continued)* Airplanes arranged by airplane manufacturer, and airport reference code.

APPENDIX C
HIGHWAY ENGINEERING

Barrier type	Description	Vehicle weight, lb	Maximum deflection, ft
Flexible			
3-strand cable	¾-in-diameter steel cables 3 to 4 in apart, mounted on weak posts spaced 12 to 16 ft	1800–4500	11.5
W-beam weak post	Similar to cable guardrail except it uses a corrugated metal rail whose cross section resembles the letter w	1800–4000	7.3
Thrie beam* weak post	Same as the weak-post, W-beam except it uses a thrie beam rail	1800–4500	6.2
Semirigid			
Box beam	Consists of a box rail mounted on steel posts (e.g., 6 in × 6-in box mounted on S3×5.7 steel posts on 6-ft centers)	1800–4000	4.8
Blocked-out W-beam (strong post)	Consists of wood or steel posts and a W-beam rail. Posts are set back or *blocked out* to minimize vehicle snagging	1800–4500	2.9
Blocked-out thrie beam* (strong post)	Same as blocked-out W-beam except with a thrie-beam rail. The added corrugation stiffens the system	1800–4000	3.3
Modified thrie beam*	Similar to a blocked-out W-beam with a triangular notch cut from the spacer block web. Minimizes vehicle rollover	Tested for 1800 lb, 20,000 lb (2.9-ft deflection), and 32,000 lb	
Self-restoring barrier (SERB) guardrail	Consists of tubular thrie beam rail supported from wood posts by steel pivot bars and cables. Classified experimental	1800–40,000	3.9
Steel-backed wood rail	Consists of wood rail backed with a steel plate and supported by timber posts	1800–4500	
Rigid			
Concrete safety shape	Similar to a concrete median barrier but has a smaller section. Has sloped front face and vertical back face	1800–4500	
Stone masonry wall	A 2-ft-high barrier consisting of a reinforced concrete core faced and capped with stone and mortar	1800–4300	

*Cross section of a thrie beam looks like three vees (vvv).

FIGURE C-1 This table outlines the standard sections of roadside barriers. *(Merrit)*

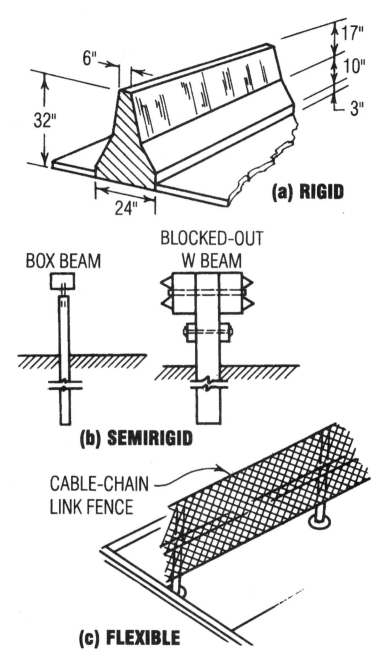

FIGURE C-2 These drawings show typical types of barriers for roadways: (a) rigid, (b) semirigid, and (c) flexible. *(Merrit)*

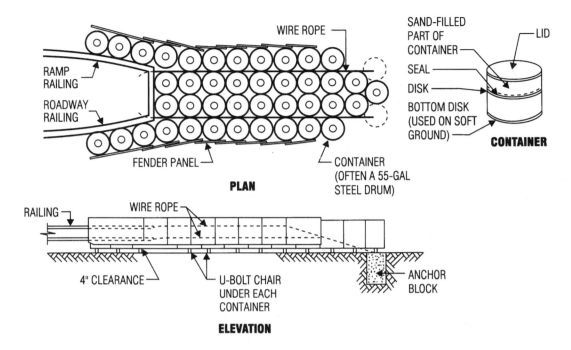

FIGURE C-3 This drawing shows containers filled with sand that are used as an inertial barrier. If a vehicle were to impact this type of crash cushion, the sand would absorb momentum from the vehicle. *(Merrit)*

Design speed, mi/h	Assumed passed-vehicle speed, mi/h*	Minimum passing sight distance, ft
30	26	1100
40	34	1500
50	41	1800
60	47	2100
65	50	2300
70	54	2500
75	56	2600
85	59	2700

*Assumed speed of passing vehicle 10 mi/h faster than that of the passed vehicle.

FIGURE C-4 This figure is a table outlining the minimum passing sight distances for designing two-lane highways. *(Merrit)*

HIGHWAY ENGINEERING **C.5**

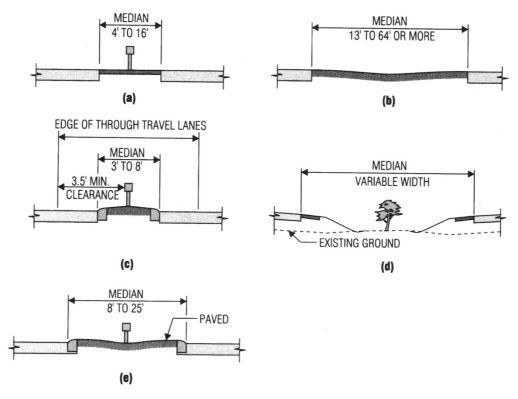

FIGURE C-5 A median is a wide strip of a highway used to separate traffic traveling in opposite directions. Different types of medians are shown here: (a) a paved flush median, (b) a median with swale and paved flush (maximum slope 1:6) when median width exceeds 36 ft.; otherwise, paved and incorporating a median barrier, (c) a raised, curbed, and crowned median, with 3-ft. width when optional median barrier is installed, (d) a median as natural ground between roadways, and (e) a median that is raised, curbed, and depressed toward the median barrier. *(Merrit)*

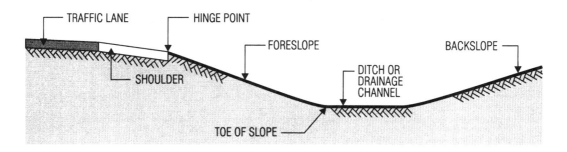

FIGURE C-6 This figure shows the typical elements of a roadside. *(Merrit)*

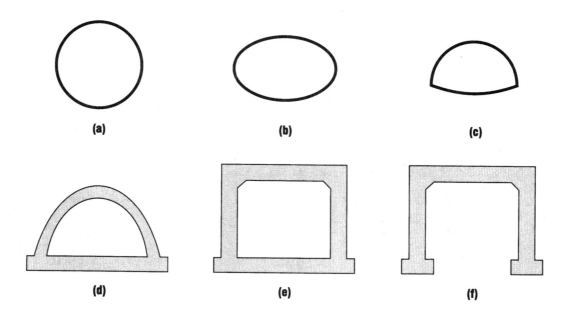

FIGURE C-7 A culvert is a closed conduit for passage of runoff from one open channel to another. These drawings show the cross-sections of different types of culverts: (a) a circular pipe, usually concrete, corrugated metal, vitrified clay, or cast iron, (b) an elliptical pipe, generally reinforced concrete or corrugated metal, (c) a precast concrete pipe arch, (d) a corrugated metal or reinforced concrete arch, (e) a reinforced concrete box culvert, and (f) a reinforced concrete bridge culvert. *(Merrit)*

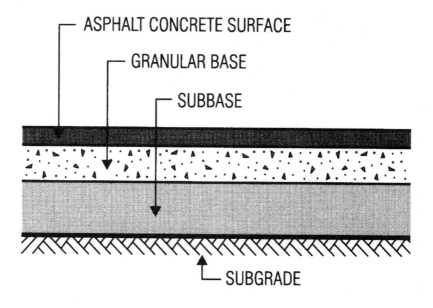

FIGURE C-8 This figure shows the main components of a typical flexible pavement. *(Merrit)*

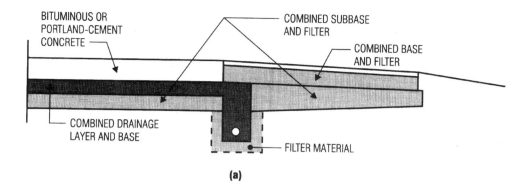

FIGURE C-9 Drainage layers under pavements are shown in this figure. The top drawing (a) shows a base used as the drainage layer, and the bottom drawing (b) shows the drainage layer as part of or below the subbase. *(Merrit)*

Traffic, ESAL	Asphalt concrete, in	Aggregate base, in
Less than 50,000	1.0[†]	4
50,000 – 150,000	2.0	4
150,001 – 500,000	2.5	4
500,001 – 2,000,000	3.0	6
2,000,001 – 7,000,000	3.5	6
Greater than 7,000,000	4.0	6

[*] Adapted from AASHTO "Guide for Design of Pavement Structures."
[†] For surface treatment.

FIGURE C-10 This table shows the minimum layer thickness (in inches) recommended by AASHTO for various levels of ESAL. *(Merrit)*

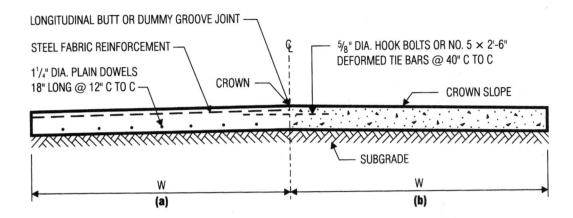

FIGURE C-11 A concrete pavement may be plain concrete, reinforced concrete, or prestressed concrete. This figure shows a cross section of a reinforced concrete pavement. On the left (a), the concrete is reinforced, and on the right (b), the concrete is not reinforced. *(Merrit)*

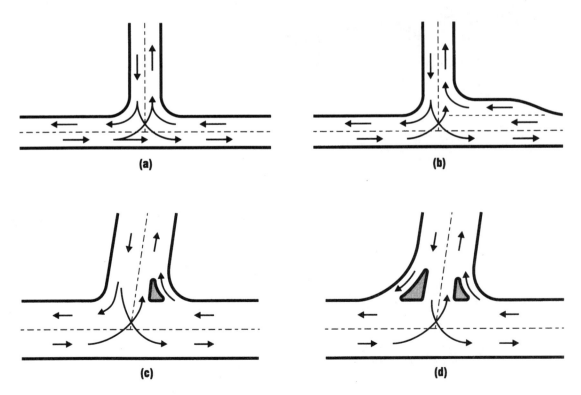

FIGURE C-12 These drawings show types of at-grade T intersections: (a) an unchannelized intersection, (b) an intersection with a right-turn lane, (c) an intersection with a single-turning roadway, and (d) a channelized intersection with a pair of turning roadways. *(Merrit)*

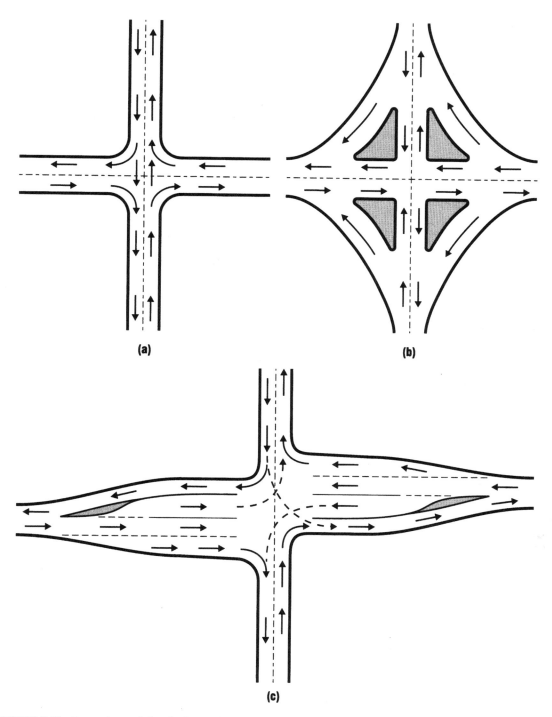

FIGURE C-13 Types of at-grade four-leg intersections are shown in these drawings: (a) an unchannelized intersection, (b) a channelized intersection, and (c) a flared intersection. *(Merrit)*

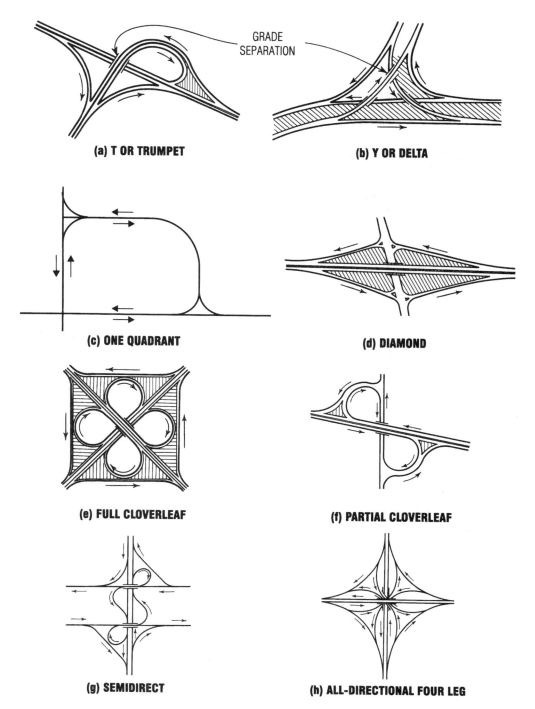

FIGURE C-14 Designers have a wide variety of interchange layouts from which to choose. These drawings show types of interchanges for intersecting grade-separated highways. *(Merrit)*

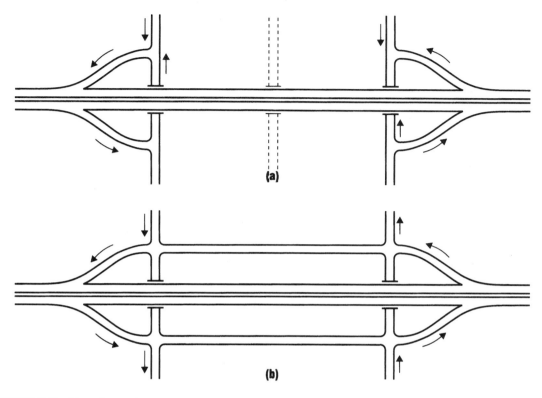

FIGURE C-15 These figures show split-diamond interchanges (a) with two-way streets and (b) with one-way streets. *(Merrit)*

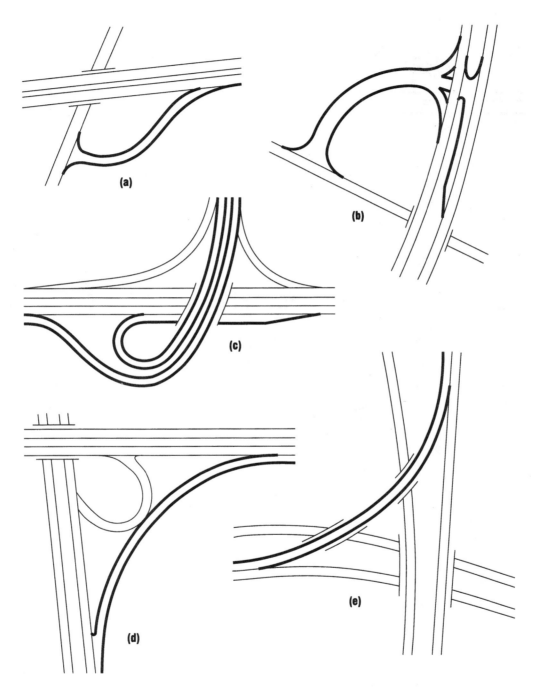

FIGURE C-16 A ramp is a roadway that connects tow or more legs of an interchange and is used for turning traffic. Types of ramps are shown here: (a) diagonal, (b) one-quadrant, (c) loop and semidirect, (d) outer connection, and (e) directional. *(Merrit)*

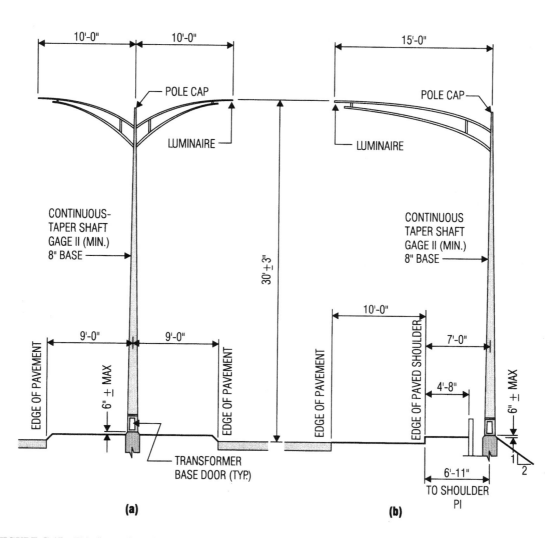

FIGURE C-17 This figure shows highway lighting installation with luminaries on tall posts. The drawing on the left (a) shows a post in the middle of a roadway with luminaries on both sides, and on the right (b), a post located on the right side of the road is shown. *(Merrit)*

INDEX

A

Additives, 11.6
Airport access, 12.1
Airport engineering, 12.1, B.1
Air pollution, 4.3
Annual increase, 6.1
Arterial capacity, 20.7
    Maximizing, 20.7
Asphalt amendments, 9.6
Asphalt cements, 11.5
Auxiliary lanes, 20.1

B

Backfill, 13.2
    Mechanically stabilized, 13.2
Base construction, 11.2
Bicycle facilities, 15.1
Bicycle streets, 15.2
Bike lanes, 15.2
Bikeways, 15.1
Borrow, 2.2
Bridge engineering, A.1
Bridges, pedestrian, 18.4
Small, 3.4
Bus transit, 16.2

C

Cantilevered walls, 13.4
Center-turning overpass, 7.5
Cloverleaf interchanges, 8.8
Commuter trains, 16.6
Concrete base, 9.6
Continuous slab, 10.4
Cross sections, 2.2
Counterfort walls, 13.4
Culverts, 3.3
    Precast, 3.4
Curbs, 3.1
Cuts, 2.2

D

Design speeds, 1.6
Design vehicles, 5.1
Design volumes, 6.1
Diamond interchanges, 8.4
Directional interchanges, 8.12
Ditches, 3.2, 5.4
Drainage, 3.1

E

Earthwork, 2.1
Easements, 19.2

Joint use of, 19.3
Permanent road, 19.3
Temporary construction of, 19.2
Edge drains, 3.2
Efficiency of operation, 17.3
Environmental protection, 4.1
Erosion, 4.2

F

Fills, 2.2
Flexible pavements, 12.4
Flexible pavement design, 11.1
Freeway design, 8.1
Freeway warrants, 8.1

G

Grade separations, 8.3
Gravity walls, 13.2
Gutters, 3.1

H

Haul, 2.3
Heavy rail, 16.4
High-speed rail, 17.4
Highway engineering, C.1
Highway-railroad crossings, 14.1
Historical sites, 4.2
Horizontal alignment, 1.1
Horizontal curves, 1.1
HOV lanes, 16.2

I

Indirect left turns, 7.5
Interchange design, 8.1
Interchange warrants, 8.1
Intersections, low-volume, 7.1
Volume-modified, 7.3
Intersection design, 7.1

L

Lane capacity, 6.2
Lanes, 5.3, 6.4
Auxiliary, 20.1
Bike, 15.2
HOV, 16.2
Light coordination, 20.5
Light rail, 16.4
Low-volume intersections, 7.1

M

MAGLEV, 17.7
Marking, 20.4
Medians, 5.4
Mix designs, 10.2
Modified diamonds, 8.6

N

Noise pollution, 4.1

P

Parking control, 20.6
Parks, 4.2
Passenger capacity, 17.3
Pavement base, 9.1
Pavement construction, 11.2
Pavement design, 10.1
Pavement reconstruction, 11.3
Pedestrian bridges, 18.4
Pedestrian facilities, 18.1
Portland cement concrete base, 9.6
Pozzolan amendments, 9.6
Prestressed concrete pavement, 10.4
Public buildings, 4.2

R

Railway engineering, 17.1
Recycled asphalt, 11.4
Reinforced earth, 13.2

Retaining walls, 13.1
    Choosing, 13.1
Right-of-ways, 19.1
    Planning, 19.1
    Scheduling, 19.2
Rigid pavements, 12.7
Rigid pavement design, 10.1
Roadway design, 5.1
Roundabouts, 7.5
Rural rail crossings, 14.1
Rubber crumbs, 11.6
Runway configurations, 12.2
Runway grades, 12.2
Runway slopes, 12.2

S

Separations, 14.4
Shoulders, 5.4
Sidewalks, 18.1
Sight distance, 1.7
Signal lights, 20.5
Signing, 20.4
Siltation, 4.2
Spiraled railroad curves, 1.4
Stone matrix asphalt, 11.7
Street crossings, 18.3
Subbase construction, 11.1
Subbase material, 9.1
Subgrade construction, 11.1
Superelevation, 1.4
Superpave mixes, 11.5

T

Tangents, 1.1
Traffic calming, 20.6
Traffic counting, 6.1
Traffic estimation, 6.1
Traffic forecasts, 6.1
Traffic management, 20.1

Transit, 16.1
Turning widths, 5.1

U

Urban congestion solutions, 15.3
Urban rail crossings, 14.3
U-turns, 7.5

V

Vertical alignment, 1.1, 1.6
Vertical curves, 1.7
Volume-modified intersections, 7.3

W

Waste, 2.2
Wetlands, 4.2
    Filling, 4.2
Wildlife habitat, 4.4

NOTES

NOTES

NOTES

NOTES

NOTES

NOTES

NOTES

NOTES

NOTES

NOTES

NOTES